T0261271

Operational Weather
Forecasting

Advancing Weather and Climate Science Series

Series Editors:
Peter Inness, University of Reading, UK
William Beasley, University of Oklahoma, USA

Other titles in the series:

Mesoscale Meteorology in Midlatitudes
Paul Markowski and Yvette Richardson, Pennsylvania State University, USA
Published: February 2010
ISBN: 978-0-470-74213-6

Thermal Physics of the Atmosphere
Maarten H.P. Ambaum, University of Reading, UK
Published: April 2010
ISBN: 978-0-470-74515-1

The Atmosphere and Ocean: A Physical Introduction, 3rd Edition
Neil C. Wells, Southampton University, UK
Published: November 2011
ISBN: 978-0-470-69469-5

Time-Series Analysis in Meteorology and Climatology: An Introduction
Claude Duchon, University of Oklahoma, USA and
Robert Hale, Colorado State University, USA
Published: January 2012
ISBN: 978-0-470-97199-4

Operational Weather Forecasting

Peter Inness
University of Reading, UK

Steve Dorling
University of East Anglia, UK

WILEY-BLACKWELL

A John Wiley & Sons, Ltd., Publication

This edition first published 2013 © 2013 by John Wiley & Sons, Ltd

Wiley-Blackwell is an imprint of John Wiley & Sons, formed by the merger of Wiley's global Scientific, Technical and Medical business with Blackwell Publishing.

Registered office: John Wiley & Sons, Ltd, The Atrium, Southern Gate, Chichester, West Sussex, PO19 8SQ, UK

Editorial offices: 9600 Garsington Road, Oxford, OX4 2DQ, UK
 The Atrium, Southern Gate, Chichester, West Sussex, PO19 8SQ, UK
 111 River Street, Hoboken, NJ 07030-5774, USA

For details of our global editorial offices, for customer services and for information about how to apply for permission to reuse the copyright material in this book please see our website at www.wiley.com/wiley-blackwell.

The right of the author to be identified as the author of this work has been asserted in accordance with the UK Copyright, Designs and Patents Act 1988.

All rights reserved. No part of this publication may be reproduced, stored in a retrieval system, or transmitted, in any form or by any means, electronic, mechanical, photocopying, recording or otherwise, except as permitted by the UK Copyright, Designs and Patents Act 1988, without the prior permission of the publisher.

Designations used by companies to distinguish their products are often claimed as trademarks. All brand names and product names used in this book are trade names, service marks, trademarks or registered trademarks of their respective owners. The publisher is not associated with any product or vendor mentioned in this book. This publication is designed to provide accurate and authoritative information in regard to the subject matter covered. It is sold on the understanding that the publisher is not engaged in rendering professional services. If professional advice or other expert assistance is required, the services of a competent professional should be sought.

Library of Congress Cataloging-in-Publication Data

Inness, Peter (Peter M.)
 Operational weather forecasting / Peter Inness and Steve Dorling.
 p. cm.
 Includes bibliographical references and index.
 Summary: "This book will cover the end-to-end process of operational weather forecasting"–Provided by publisher.
 ISBN 978-0-470-71159-0 (cloth) – ISBN 978-0-470-71158-3 (pbk.) 1. Weather forecasting. I. Dorling, Steve (Stephen R.) II. Title.
 QC995.I44 2013
 551.63–dc23

 2012025403

A catalogue record for this book is available from the British Library.

Wiley also publishes its books in a variety of electronic formats. Some content that appears in print may not be available in electronic books.

Set in 11/13pt Palatino-Roman by Laserwords Private Limited, Chennai, India

1 2013

Contents

Series Foreword

Advancing Weather and Climate Science

Meteorology is a rapidly moving science. New developments in weather fore-casting, climate science and observing techniques are happening all the time, as shown by the wealth of papers published in the various meteorological journals. Often these developments take many years to make it into academic textbooks, by which time the science itself has moved on. At the same time, the underpinning principles of atmospheric science are well understood but could be brought up to date in the light of the ever increasing volume of new and exciting observations and the underlying patterns of climate change that may affect so many aspects of weather and the climate system.

In this series, the Royal Meteorological Society, in conjunction with Wiley-Blackwell, is aiming to bring together both the underpinning principles and new developments in the science into a unified set of books suitable for undergraduate and postgraduate study as well as being a useful resource for the professional meteorologist or Earth system scientist. New developments in weather and climate sciences will be described together with a comprehensive survey of the underpinning principles, thoroughly updated for the 21st century. The series will build into a comprehensive teaching resource for the growing number of courses in weather and climate science at undergraduate and postgraduate level.

Series Editors

Peter Inness
University of Reading, UK

William Beasley
University of Oklahoma, USA

Preface

If you are interested in learning how weather forecasts are produced in today's forecasting centres then this book aims to be a complete primer, covering the end-to-end process of forecast production. Other textbooks cover specific aspects of the process and, in particular, the formulation of numerical models, but here we aim to bring a description of all the relevant aspects together in a single volume, with plenty of explanation of some of the more complex issues and examples of current, state-of-the-art practices.

This book grew out of a module on 'Operational Forecasting Systems and Applications', which is part of the University of Reading MSc in Applied Meteorology. Because the University of Reading also runs an MSc course in Numerical Modelling of the Atmosphere and Oceans, and another in Data Assimilation, the module deliberately avoids too much detail on the mathematical formulation of numerical models and the statistical and numerical formulation of data assimilation schemes. This book follows the same approach and is intended to be an overview of the end-to-end process of weather forecast production at a major National Weather Service. The physics and numerics of models and the formulation of data assimilation schemes are touched upon in the module, but other options exist for students at Reading if they want to learn about these aspects in more detail. Because the students at Reading come from all around the world, the module also attempts to be as generic as possible when discussing operational weather forecasting. However, because of our location in the United Kingdom (with UK Met Office staff based in the University's Meteorology Department) and with the European Centre for Medium-range Weather Forecasts (ECMWF) two miles down the road, many of the examples included in the module (and hence this book) are based on practices at these two major forecasting centres.

When I started to put together material for the MSc module in 2005, it soon became clear that there was no good single textbook available on the process of operational weather forecast production. There were several excellent texts on the numerical formulation of models and the design of data assimilation schemes, and several more have appeared since. The lack of books on operational weather forecasting is probably because this is a rapidly developing field, with a lot of variations in practices between different forecasting centres, making its operational application a difficult field to describe in a textbook. Another issue is that the people involved in operational forecasting on a day-to-day basis are also far too busy doing their jobs to take time out and write a book about it all! I discovered this when, as series editor for this Wiley-Blackwell series on Advancing Weather and Climate Science, I started to approach potential authors for a textbook on this subject. Many people said that a textbook would be useful but they were unable to commit themselves to writing one.

I decided, therefore, that it would be worth trying to commit the material from the Reading MSc module, together with further detail and examples, to a book. Steve Dorling at the University of East Anglia, who teaches similar modules while also being Innovations Director at the UEA-based Weatherquest Ltd, a private-sector company, agreed to join me in this venture. We realised at the outset that we would have to accept that some aspects of the book would become out-of-date fairly rapidly, and that using specific examples of practices at one forecasting centre might also alienate potential readers with an interest in practices at a different centre. Despite this we agreed that it was worth trying to commit the current state of the art to a textbook and have tried to be as generic as possible when describing aspects of numerical model design and formulation. We hope the final product and any subsequent editions will be adopted by professional meteorological training colleges, by universities and indeed by individuals fascinated by meteorology.

This book then is aimed primarily at advanced undergraduate and masters level students of meteorology and so some knowledge of basic meteorology is assumed. Many universities around the world teach courses in theoretical meteorology and numerical modelling, using the many excellent books that are available covering these topics. Our book is aimed at helping the students on those courses to understand how the theory comes together in the day-to-day applications of weather forecast production. To some extent it could be regarded as an aid to converting from a student of meteorology into an operational practitioner. Discussions with staff in operational forecasting centres who are involved in the recruitment and training of new staff have commented that this would be a very useful purpose for a textbook. Many students leave university courses with an excellent grasp of the theory of meteorology but only a rather slim knowledge of the actual practice of forecast production.

This book is neither a manual on how to build a numerical model or how to be an operational weather forecaster but instead aims to cover the whole process of forecast production, from understanding the nature of the forecasting problem, gathering the observational data with which to initialise and verify forecasts, designing and building a model (or models) to advance those initial conditions forwards in time and then interpreting the model output and putting it into a form which is relevant to customers of weather forecasts. One subject which we wanted to include was the generation of forecasts on the monthly-to-seasonal timescales. This has been an area of research for many years and some operational centres have been doing this type of forecasting for some time but again it is not well covered in textbooks.

As far as references are concerned we have taken a fairly sparse approach. When describing fundamentals of numerical weather prediction (NWP) model design we have referenced a very few seminal papers but other textbooks that cover this topic in more detail (such as Eugenia Kalnay's excellent 'Atmospheric Modeling, Data Assimilation and Predictability') already provide extremely comprehensive reference lists in this area. We have included references when describing specific schemes, methods and techniques used at different operational forecasting centres, or particular studies conducted at these centres. However, because operational forecasting systems change rapidly and undergo frequent upgrades, many new developments never make it into the wider literature before they become out of date. The priority of operational centres is to maintain and develop their forecasting systems and update their internal documentation rather than publish their methods in the reviewed literature. Much of the detail of operational forecasting products in this book also comes from web sites which change regularly. We have provided web addresses where appropriate but the reader must be prepared for the information on those sites to change or for the addresses themselves to change or disappear.

Peter Inness
University of Reading

Acknowledgements

A number of forecasting centres have given us access to their documentation, training material and staff to help in the production of this book. In particular we would like to thank the European Centre for Medium-range Weather Forecasts (ECMWF) for a large amount of material from its excellent training courses which appears throughout the book. The UK Met Office has also provided a considerable amount of material and we would particularly like to thank Tim Hewson, a senior forecaster at the UK Met Office, for his guided tour around the National Meteorological Centre at Exeter and discussions on the role of a senior forecaster within a large forecasting organisation. The NOAA Climate Prediction Center and the online plotting and analysis facility of the NOAA Earth Systems Research Laboratory have also proved to be excellent sources of data and figures.

I would like to thank many colleagues at the University of Reading for interesting discussions on many aspects of theoretical and applied meteorology which have helped build the framework for this book. Thanks also to former colleagues at the Met Office College (1996–1999) with whom I worked training forecasters and meteorological research staff. The things I learnt whilst working there have been invaluable in many aspects of my current work. Thanks are also due to all the University of Reading students who have provided feedback on the MSc module 'Operational Forecasting Systems and Applications'. Much of this feedback has been incorporated into this book – not least the suggestion that it ought to be written in the first place. Finally, thanks to my family who keep me from getting too absorbed in the work and help me relax and enjoy life.

SD would like to thank the University of East Anglia (UEA) for granting a sabbatical period which was very valuable in preparing material for the text. Thanks also to ECMWF for the stimulating course on 'Use and Interpretation of ECMWF Products', to Weatherquest Ltd colleagues and clients for the chance to discuss, design and deliver operational services, and to all the students of meteorology at UEA for their enthusiasm over the last twenty years. To Tim, Val, Rachel, Heather, Lewis and Lyndsey Dorling, thank you for the greatest support of all.

1

Introduction

The production of weather forecasts for use by the general public, governments, the military, news media and a wide range of industrial and
commercial activities is a major international activity. It involves tens of
thousands of people and many billions of pounds worth of high-tech equipment, such as computing and telecommunications networks, satellites and
land-based observing systems. Most nations across the world have a national
meteorological agency which generates forecasts for both government and
commercial customers, although the size and scope of these agencies varies
widely from country to country. There are also many private companies
producing weather forecasts. These companies often specialise in forecasts
for particular niche markets – news media, commercial shipping, offshore oil
and gas exploration and production, the agricultural sector and so on. In
some cases, weather forecasts can have very large financial benefits to particular customers. Power generation companies can make many thousands of
pounds on the basis of a single weather forecast of a cold spell, as this allows
them to buy gas at a low price prior to the increased demand during cold
weather pushing the price up. Offshore gas and oil exploration companies
can avoid multimillion dollar damage to drilling platforms by shutting down
operations prior to the onset of a severe storm. Major supermarket chains use
weather forecasts to plan their stock control in the knowledge that a spell of
cold weather will result in increased sales of foods like soup whereas a spell
of warm weather will increase demand for ice cream and barbecues.

The production of weather forecasts is based on a sound scientific understanding of how the atmosphere works, coupled with a vast amount of
technical investment in observing, computing and communication facilities.
The current state-of-the-art forecasting facilities at the world's major meteorological centres have evolved over a period of many years and are the
result of scientific research and technical development. At the heart of any
major weather forecasting centre is a numerical weather prediction (NWP)

Operational Weather Forecasting, First Edition. Peter Inness and Steve Dorling.
© 2013 John Wiley & Sons, Ltd. Published 2013 by John Wiley & Sons, Ltd.

model – a computerised model of the atmosphere which, given an initial state for the atmosphere, derived from observations, can generate forecasts of how the weather will evolve into the future. Prior to the advent of NWP, forecasts were produced manually, usually by forecasters drawing maps of the current state of the atmosphere and then, using their knowledge of typical weather patterns together with a set of empirical rules, attempting to predict how that state would evolve. These days the computer model has taken over the task of predicting the evolution of the atmospheric state but there is still a big role for the human forecaster in the system.

This book is not a guide to the detailed formulation and coding of an NWP model. Nor is this book a manual on how to be a weather forecaster. Individual meteorological services produce training manuals for their forecasters which go into the specifics of forecast production within their organisations. Rather, this book will describe the end-to-end process of weather forecast production, with a focus on the NWP tools available to forecasters in major meteorological centres. The nature of the weather forecasting problem is discussed in Chapter 2. Chapter 3 focuses on the observations of the weather that allow forecasting centres to set the initial conditions for their forecasts. These same observations also form an important part of the forecast production process in the sense that human forecasters will continually check the observations against the NWP forecasts and modify the forecasts as and when necessary. Chapters 4 and 5 look at the basic ingredients of NWP models and how these ingredients are applied in order to produce operational forecasting systems for specific tasks. In Chapter 6 the role of the human forecaster within the NWP forecast production process is examined. Chapter 7 concentrates on the specific challenge of forecasting for periods of several weeks to several months ahead and, finally, Chapter 8 looks at how NWP forecasts are verified and measured, and how this process then feeds into the process of continuing development and improvement.

1.1   A brief history of operational weather forecasting

Many different human activities have always been sensitive to prevailing weather conditions and so people have been trying to predict the weather, both on a day-to-day basis and for the coming few months, since ancient times. As far back as the earliest civilizations, the weather during the growing season has affected crop production and so farmers have always taken note of the weather and climate. The ancient Egyptians, for instance, kept detailed records of the flooding of the Nile – an annual event that had a big impact on soil fertility and which was strongly affected by the intensity of the rainy season around the headwaters of the Nile.

Any kind of weather prediction prior to the scientific advances of the nineteenth century was largely based on folklore and perceived relationships between events in nature and the weather. Certainly there was no scientific basis to most of these forecasting methods but farmers and people who spent most of their days in the open air were certainly well attuned to the prevailing conditions and by observing changes in clouds and wind could make reasonably good predictions of the forthcoming weather for the next few hours or even a day or two. However, there was no systematic attempt to predict the weather in any kind of organised way.

The first organised meteorological agency was set up by the British government in the mid-nineteenth century as a response to loss of shipping on the trade routes that sustained the British Empire. In 1854 the British Board of Trade appointed Admiral Robert FitzRoy as its 'meteorological statist'. FitzRoy was an oceanographic surveyor with a reputation for producing detailed and accurate hydrological charts of coastal waters around the world. The Board of Trade hoped that he would also be able to produce an equivalent *meteorological* atlas charting weather conditions around the world which would serve to inform shipping lines and crews of the risks of storms, allowing them to make choices about when to sail and what routes to take.

FitzRoy began this task but he was also interested in the possibility of being able to issue more specific predictions of the weather conditions in coastal waters around Britain on a day-to-day basis. Since the invention of the barometer by Toricelli in the seventeenth century, people had started to realise that atmospheric pressure was correlated with weather conditions, with falling pressure often presaging unsettled or even stormy conditions, and rising or steady pressure indicating settled conditions. FitzRoy made use of this concept by designing a barometer specifically for use in ports, with information on what weather conditions to expect given an observed trend in the pressure. A version of this barometer was set up in all the major harbours around the United Kingdom, so that captains could consult the barometer prior to setting sail. Interestingly, as well as advice on the expected weather conditions associated with rising and falling pressure, there was also advice on how to interpret the colour of the sky at dawn and dusk in terms of forthcoming weather conditions. The recent advent of the electric telegraph allowed FitzRoy to take the use of these barometers one stage further. Regular readings from the barometers around the country, together with information about other weather variables, such as wind and cloud, could be telegraphed back to FitzRoy's office in London and the readings could be plotted onto a chart that summarized the atmospheric pressure distribution around the country. A sequence of these charts could then be used to predict the winds and prevailing weather around the country in the coming hours and even a day or so ahead. Furthermore, by building up an archive of these charts – today we'd probably call it a database – it might be possible to compare the current pressure distribution with similar pressure patterns from the past and use the knowledge of what happened to the weather in the earlier cases to predict what might happen this time. Predictions could then be telegraphed back to the port authorities. In this way FitzRoy invented the concept of operational weather forecasting, and probably also coined the term 'weather forecast'.

Of course, the infinitely variable nature of the weather meant that many of FitzRoy's forecasts went wrong and, particularly after making his forecasts more widely available to the public through the daily newspapers, FitzRoy received a lot of criticism from the public and the scientific community. This criticism was one of the several factors that led to FitzRoy's suicide in 1865. The work of the Meteorological Office of the Board of Trade continued though, with the forecasting methods being refined and new ones being developed. Napier Shaw took over the directorship of the Meteorological Office in 1905 and was instrumental in introducing more scientifically based forecasting methods.

The French government set up its first national weather service in 1855, once again as a response to the loss of shipping – this time due to a storm in the Black Sea during the Crimean War. In 1870 the US government, under President Ulysses S. Grant, also set up a meteorological service, this time under the auspices of the US War Office. Grant's recent experience as a general in the American Civil War had made him well aware of the impact of the weather on military operations, and the military forts around the United States provided the ideal locations for a weather observing network. The US National Weather Service became a civilian agency in 1890 when it was moved into the Department of Agriculture. The Australian Bureau of Meteorology was set up in 1906, although prior to this date each state had its own meteorological service.

Throughout the twentieth century, further scientific advances in weather prediction continued to be made. The British meteorologist Lewis Fry Richardson developed, single-handed, a method of forecasting the evolution of the state of the atmosphere using the set of physical equations that govern atmospheric motion. During World War I, in which he served as an ambulance driver, Richardson developed a way of solving these equations numerically and even performed a six-hour forecast of the pressure in central Europe, using data from 20 May 1910. This was a Herculean task involving many thousands of calculations that needed to be performed and double checked without the aid of any kind of electronic calculation device. The result was completely wrong but in his 1922 book, *Weather Prediction by Numerical Process*, Richardson set down his methods; these have formed the basis of numerical weather prediction ever since.

During World War II, the massive use of military aviation and shipping to conduct the fighting over wide areas demanded advances in weather forecasting. Aircraft on long range bombing missions relied on favourable weather conditions in order to be able to see the ground on arrival over their targets and naval vessels, particularly those involved in the Pacific campaign in the part of the world most prone to tropical cyclones, needed to be able to avoid damaging storm conditions. Weather observing and forecasting practices improved rapidly through this period, with forecasters for the first time starting to pay attention to the development of the middle and upper troposphere in order to determine what might happen to the weather. The use of weather balloons carrying instrumented packages to observe the upper troposphere became more widespread as a result. Probably the most famous single weather forecast of all time was made during WWII, with forecasters led by Captain James Stagg advising the Allied Command to delay the D-Day Normandy landings in June 1944 by 24 hours.

Following the war, research into forecasting methods followed two main strands. In the United Kingdom meteorologists developed techniques based on looking at maps of the state of the middle and upper troposphere in order to predict where significant weather system developments would occur. In the United States researchers at a number of different institutions were working on developing and refining the numerical methods proposed by Richardson to produce forecasts. The availability of numerical computing devices, such as the ENIAC machine, greatly aided this strand of development. The results of the first computerised atmospheric forecast were published in 1950 by a research group at Princeton University, which included the mathematician John von Neumann and the meteorologist Jule Charney. The numerical model used had been developed over the preceding few years and was actually somewhat simpler than that used by Richardson. However, it produced a reasonably good 24-hour prediction of the evolution of the mid-tropospheric flow over the continental United States and this encouraging result led to further refinements of the numerical methods. At this stage, the method was in no way ready for operational use – the computing time alone was about 24 hours for a 24-hour forecast and this didn't include the time spent preparing the initial conditions for the forecast and inputting them into the computer.

The first *operational* numerical weather forecasts were made by the Swedish Military Weather Service in 1954. The development of the methods used in these forecasts was led by Carl-Gustav Rossby, a native Swedish meteorologist who had worked in the USA, principally at Chicago University, during the 1930s and 1940s. On returning to Sweden in 1947 he founded the Swedish Institute of Meteorology and continued to develop numerical forecasting methods.

The UK Met Office came to numerical weather prediction (NWP) rather late. During the 1950s and early 1960s the UK Met Office had to do its research into NWP on borrowed computers and it wasn't until the mid-1960s that it actually had its own computing facility. In 1965 the UK Met Office started to routinely produce numerical weather forecasts and in 1967 it had the distinction of producing the first numerical prediction of precipitation. Prior to this, numerical forecasts had only predicted the evolution of pressure, geopotential height and vorticity patterns and it was left to experienced forecasters to interpret these patterns in terms of the actual weather that would occur in association with them.

By the 1970s most of the major meteorological agencies around the world were well established and starting to use NWP methods as the basis of their operational forecasts. One major development in the 1970s was the setting up of the European Centre for Medium-range Weather Forecasts (ECMWF), in Reading, UK, in 1975. The aim of the centre was to develop operational NWP forecasts for the medium range (out to about 15 days) using funding from all the main European meteorological services, which would not have been able to develop such facilities with their own resources. The forecasts would then be made available to the National Weather Services of all the member states. ECMWF pioneered the operational use of *ensemble forecasting* techniques (more details are given in Chapter 5) and has since developed numerical methods and models for prediction on the monthly to seasonal timescale. Such techniques require vast amounts of computing power and it is only through the collaborative funding of all the member states that such facilities can be maintained and upgraded. The supercomputing facilities at ECMWF regularly top the league table of computing power in the United Kingdom and the NWP forecasts produced using these facilities are widely regarded as being the best in the world. The US National Center for Environmental Prediction (NCEP) started to produce ensemble forecasts in the early 1990s. More recently many forecasting centres have introduced ensemble methods as computing facilities have developed. The UK Met Office and the Australian, Chinese, Japanese, Korean, French, Brazilian and Canadian National Meteorological Services all now routinely run ensemble forecasts in some form.

Since the 1970's NWP methods have become more accurate and the models used have become faster. Communication networks have improved massively over this period too. As recently as the early 1980's an outstation weather forecaster working at a military airfield for instance, would have had very little access to the output from numerical models, and what few products were disseminated often arrived too late to be of any use in making forecasts for the aircrews. This lack of up-to-date model output made many forecasters somewhat wary of using NWP products to guide their forecasts. Similarly outstation forecasters saw very few satellite images. It wasn't until the advent of high bandwidth communication networks in the 1990's that forecasters working at locations remote from the main meteorological service headquarters got to see a wide range of guidance from numerical models together with

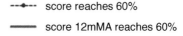

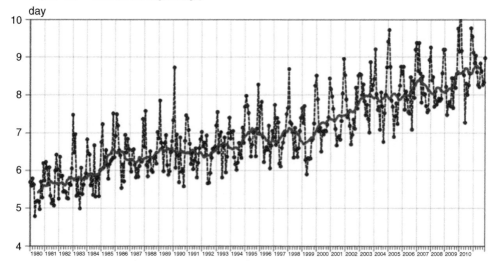

Figure 1.1   A time series of the anomaly correlation between the forecast and observed 500 hPa geopotential height anomaly in the northern hemisphere extra-tropics from the ECMWF forecast model. The vertical scale shows the point of the forecast in days at which this correlation falls below 60%. The blue dashed line shows this value for every month and the red solid line shows the 12-month running mean. (Reproduced by permission of ECMWF.)

regular, detailed satellite imagery. So going right back to FitzRoy who was able to make use of the newly invented electric telegraph system, it is clear that communication networks play a vital part in operational meteorology.

Improvements in forecast accuracy since the 1980s are illustrated in Figure 1.1. It shows, for the ECMWF model, the point in the forecast at which the correlation between the forecast and observed 500 hPa geopotential height anomaly in the Northern hemisphere extra-tropics falls below 60%. This value is considered to be a measure of the usefulness of the forecast, with values above 60% representing skilful forecasts. The blue dashed line shows the value for each month since January 1980, and the red solid line shows the 12-month running mean of the monthly values. In 1980 the correlation fell below 60% at about 5.5 days into the forecast. By 2010 this had risen to about 8.5 days into the forecast. Effectively this means that in 2010 ECWMF was producing forecasts which were skilful for an average of three days longer than in 1980. Other forecast centres show similar improvements in skill. Of course, to most users of weather forecasts the anomaly correlation of the Northern hemisphere 500 hPa geopotential height is a pretty meaningless measure of the quality of a weather forecast and there are many ways of measuring forecast skill that are more focused on the interests of specific customers. These are discussed in detail in Chapter 8.

Recent developments in operational forecasting have been many and varied, all helped along by regular increases in the availability of computing power to meteorological agencies. More forecasting centres are now running ensemble forecasting systems and a wider range of numerical models, some with global coverage and others with very fine scale resolution over limited areas. So-called 'storm resolving models', which can explicitly represent organised convective storms, have started to come into operational use over the past few years and new ways of incorporating observations, such as rainfall rates from meteorological radar, into models are improving the way that these highly detailed forecasts are initialised. At the other end of the time and space scales, more forecasting centres are now running monthly and seasonal forecast models. Some of these developments are described in more detail in the relevant chapters of this book.

Despite the advances in technology which are driving improvements in forecast accuracy, there is still a fundamental place in the weather forecasting process for human experts. Even the most sophisticated numerical models of the atmosphere may produce forecasts which diverge significantly from reality even at quite short time ranges. In these circumstances a team of expert forecasters can spot the problems at an early stage and consider how best to amend the forecast. Models also have known systematic errors and biases, particularly when producing forecasts of local detail, and an experienced forecaster will be able to take account of these issues when producing forecasts for specific customers. Many customers of weather forecasts also require a human forecaster to act as an interface between them and the numerical weather forecast, such as military aircrew receiving a face-to-face briefing from a forecaster prior to flying weather sensitive missions or local government agencies needing briefing and hour-by-hour advice on potential disruption due to snow and ice or flooding. Often the most visible weather forecasters to the general public are those presenting weather forecasts on television but it must be remembered that these people are just the visible face of a large team of experts running, monitoring and analysing the numerical forecasts and deciding what the key weather issues for each day's forecast will be.

2

The Nature of the Weather Forecasting Problem

2.1 Atmospheric predictability

To solve any scientific problem, it is essential to fully understand the nature of the problem itself. The critical factors that affect the outcome need to be identified and fully integrated into the methods used to solve the problem. Weather forecasting is no exception to this. In essence, forecasting the weather is a problem of atmospheric physics, with many different physical processes contributing to the final outcome of the forecast. These processes need to be included in our forecasting methods as realistically as possible if we are to achieve skilful forecasts on a regular basis.

Weather forecasting can be thought of in terms of a mathematical problem too, and this is where real insight can be gained into how best to go about solving it. Weather forecasting, at least on the timescale of a few hours to a week or so, can be described as an 'Initial Value Problem' (IVP). Mathematically speaking, this is a problem in which the outcome is significantly determined by the conditions fed in at the start. In the case of weather forecasting, these conditions would be the state of the atmosphere at the starting point of the forecast. If we wish to produce a forecast for tomorrow, we need to start by determining, as precisely as possible, the state of the atmosphere today. We can then apply the various physical laws to those conditions to advance the state of the atmosphere forward in time towards tomorrow's forecast state. If instead we start the forecast with *yesterday's* weather as our initial condition, and have a perfect forecasting system which takes account of every possible relevant physical process, then using this system to advance the state of the atmosphere forwards by 24 hours will end up giving us a forecast of today's weather, not tomorrow's.

It seems obvious, therefore, that accurately specifying the initial conditions for a weather forecast is as crucial to the outcome of the forecast as

Operational Weather Forecasting, First Edition. Peter Inness and Steve Dorling.
© 2013 John Wiley & Sons, Ltd. Published 2013 by John Wiley & Sons, Ltd.

understanding all the processes that lead to the atmosphere changing its state. This has been known for many years and the art of *meteorological analysis*, whereby a skilled forecaster draws up a map or set of maps of the current state of the atmosphere (or at least its most recently observed state), has long formed a central part of the forecast process. However, it has only been since the advent of numerical weather prediction that it has become clear just how critical the accurate specification of the initial conditions of a weather forecast actually is to the outcome of that forecast. The pioneering work in this area of Ed Lorenz and others during the 1950s and 1960s has passed into meteorological legend. Meteorologists and mathematicians at this time were interested in the non-linear nature of the equations which govern the evolution of the state of the atmosphere and many other physical systems (Box 2.1).

Box 2.1 What is a non-linear system?

The easiest way to describe the characteristics of a non-linear system is to first describe a linear one. A linear system is one in which, if a change to the initial conditions produces a change to the state of the system of size x at some time t in the future, then if we multiply the change to the initial conditions by a factor a then the change to the state of the system at time t will be proportional to a. So a car driving at a constant speed and taking a certain length of time to complete a journey is a linear system. If the car drives at twice the speed, the journey will take half the time. If the car travels at half the speed the journey will take twice as long. A non-linear system is one in which this proportionality does not apply. Non-linear systems contain feedbacks which can amplify or dampen initial changes to the state of the system. Very few real physical systems are actually linear.

Lorenz demonstrated the fact that small changes to the state of the atmosphere in a numerical model can result in large changes in the evolution of that state by careful mathematical analysis of the equations that govern atmospheric motion, and also by analysing other simpler but still non-linear systems. However, the example that really brings home the effect of the non-linearity of the atmosphere happened almost by mistake. Lorenz wanted to re-run a segment of a numerical forecast rather than start from the very beginning of the forecast in order to save computing time. He re-started his new forecast using numbers taken from a paper printout that were given to three decimal places. However, the computer hardware was using numbers to six decimal places. He found that his new forecast soon started to diverge significantly from the original despite the fact that the difference in the fourth and subsequent decimal places between the new initial conditions and the original numbers was actually far smaller than the accuracy to which the meteorological variables could be measured.

One consequence of this realisation is that a tiny perturbation to the atmosphere could result in a very different evolution of the weather systems around it. The most often quoted example of this is that a butterfly flapping its wings in Brazil could cause a tornado in Texas (an example taken from the title of a talk given by Lorenz himself in 1972). Of course, it is impossible to perform an experiment with the real atmosphere in which a perturbed and unperturbed atmosphere can be compared to see how differently they evolve, but it is accepted as fact that small perturbations in a complex fluid system such as the atmosphere can indeed significantly affect the subsequent evolution of that system.

This tells us something quite profound about the nature of weather fore-casting. However comprehensively we observe the state of the atmosphere at the initial time of a forecast, the instrument errors inherent in our measuring systems alone will mean that the forecast will start to diverge from reality as the forecast evolves in time. And, of course, the meteorological observations that we do have by no means give a comprehensive coverage of the entire atmosphere, as we shall see in Chapter 3. This means that every numerical weather forecast that will ever be made is doomed to become wrong at some point. Add to this the facts that the equations which we are using to predict the evolution of the atmosphere and the methods we use to advance the forecast forwards in time are also less than perfect and one starts to marvel, as Lorenz himself did, that any weather forecasting at all is even possible. However, on a day-to-day basis, it is clear that weather forecasts for the next few hours to days ahead can indeed demonstrate useful skill although there are times when they do go quite wrong, even at rather short lead times.

So every weather forecast will start to diverge from reality. Put another way, the atmosphere is inherently *unpredictable*. It would be nice to know at what point the divergence of the forecast from reality will start to become significant but this, too, is unpredictable and is itself dependent on the state of the atmosphere. In some conditions weather forecasts a week or more ahead turn out to be rather good, but in other conditions the forecast just one day ahead can be quite inaccurate. Forecasts for a location in the Middle East during July for instance can be extremely accurate perhaps 10 days ahead due to the very small variations in the weather in that part of the world during the summer. However, forecasts for somewhere like Iceland in November can be very wrong just one day ahead due to the strong variability in weather conditions over the North Atlantic on very short timescales.

Many studies have examined the limits of atmospheric predictability and the general consensus is that something around 14 days represents a limit for *deterministic* forecasts. What people usually mean by a 'deterministic' forecast is one that gives specific values for meteorological variables such as temperature, windspeed, rainfall amount and duration and so on. Strictly speaking, however, a deterministic forecast is one in which the future state of the system is predicted by extrapolating the current state forwards in time using a fixed law or set of laws. The 14-day limit will generally be lowered by the fact that the numerical models of the atmosphere we use for prediction are themselves imperfect and so 7–10 days is a more realistic estimate of the limits of predictability. This may itself be an overestimate in situations where the atmosphere is undergoing rapid changes and large fluctuations.

Forecasters would like to have a feel for how predictable the atmosphere is every time they issue a new forecast to their customers. They could express this knowledge in terms of confidence in their forecast or indeed limit the lead time of their forecast to be shorter than the predictability limit for that particular initial state of the atmosphere. However, the predictability of the atmosphere is itself unpredictable – it is flow dependent as well as depending on the quality of the observations used to set the initial conditions.

One approach to this problem, which is described in much more detail in Chapter 5, is the use of *ensemble forecasts*. Instead of running a numerical weather forecast once, from a single set of initial atmospheric conditions, an ensemble forecast involves running the forecast 'many' times, each with a set of initial conditions which are slightly perturbed from the original, but all equally likely given the available observations and their known errors. If all of the many forecasts produce a very similar evolution of the atmosphere through the entire forecast period then it can be assumed that the atmosphere is in a state which is not sensitive to small perturbations in the initial conditions and we can be relatively confident in the forecast. If, however, all the forecasts start to diverge rapidly from each other only a short time into the forecast period then we can assume that the atmosphere is in a rather unpredictable state and so express low confidence in the forecast or simply not issue a forecast at all beyond the point at which the forecasts start to diverge significantly. There is no way of knowing in advance which *if any* of the many forecasts will be the most realistic and it is quite possible for the atmosphere to evolve in a way which *none* of the individual forecasts (usually referred to as *ensemble members*) predicted. Hence, ensemble forecasting does not necessarily lead to more accurate forecasts but it is a method which allows forecasters to make an assessment of the predictability of the atmosphere each time they issue a new forecast. Another benefit of ensemble forecasting is that it allows the generation of probability forecasts, whereby instead of making a deterministic forecast of a particular event (e.g. 'It will rain in London tomorrow morning') a forecaster can issue a forecast of the probability of that event happening (e.g. 'There is an 85% chance of rain in London tomorrow morning'). This might sound like hedging one's bets as a forecaster but probability forecasts have a real economic benefit to certain types of customer, as shown in Chapter 8.

2.2   The importance of observations in weather forecasting

It hardly needs to be stated that the availability of good quality observations of the state of the atmosphere, with appropriate temporal and spatial coverage, is crucial to the generation of skilful weather forecasts. The discussion above has focused upon the critical sensitivity of numerical weather forecasts to the specification of the initial conditions. It has also become clear, though, that no matter how good our observations are and however comprehensive their coverage is, it will still never be possible to produce perfect deterministic weather forecasts out to lead times of more than a couple of weeks.

The importance of good quality observations to the weather forecasting process is often most apparent in situations where weather conditions are changing rapidly and the atmosphere is in a 'mobile' state. The way that the initial conditions for a numerical weather forecast are set is discussed in detail in Chapters 4 and 5. But, put very simply, the initial conditions are set using a blend of the short-range forecast from the previous run of the computer model (known as the 'background field') and a combination of many different types of measurement. In situations where the state of the atmosphere isn't changing rapidly therefore, most of the important information will already be contained in the background field and the observations will only be making small adjustments to this state. A study in 2004 by Cardinali *et al.*, using the European Centre for Medium-range Weather Forecasts (ECMWF) operational forecast model, looked at the relative amounts of information coming from the observations and the background field in the initial conditions of NWP forecasts. They showed that, on average, over the period of boreal Spring 2003, the amount of information contained within the initial conditions which was due to the observations was only 15%, with the other 85% coming from the background field. Of course, the background field is itself influenced by observations inserted during the previous runs of the model, so this 15% estimate is probably a lower bound estimate. Even so, this perhaps implies that the observations are of rather secondary importance. This is not true of course, as the 15% of the information which comes from the observations could be regarded as the most crucial 15%; the information we didn't know about prior to the new model run. If we never used any observations whatsoever and just started each numerical weather forecast from the background field then our forecasts would rapidly become useless. In fact, this would be exactly the same as just running one continuous forecast that never took account of any observations of the atmosphere, so after 7–10 days of forecast time (possibly sooner) we would run into the predictability limit.

The real importance of good quality observations becomes very clear in rapidly evolving situations when it is possible that the information coming from the observations might considerably exceed 15%. An excellent example of this comes from another study made using the ECMWF model by Leutbecher *et al.* in 2002. This study looked at the damaging depression that swept across Europe on the 26 December 1999 – a storm named 'Lothar' by the Free University of Berlin. This storm caused extensive damage to buildings, electricity and telephone networks and forests across a broad swathe of France and Southern Germany; it led to 137 deaths. The costs of this storm were estimated to exceed US\$ 10 billion. The low pressure centre developed initially on the western side of the Atlantic and made a rapid crossing of the ocean. It underwent rapid intensification off the west coast of France prior to making landfall and then tracked across Northern France and central Germany. Several operational forecast models failed to forecast its track, intensity and development at quite short lead times (1–3 days). Figure 2.1 shows the forecast position of the Lothar storm from the ECMWF model at a lead time of 48 hours, together with the actual position of the storm at this time. The forecast predicted a depression centred slightly to the southeast of its actual position, but this was not deep enough by 15 hPa. The result of this was that the predicted winds across Northern France and Germany were too weak in the forecast. The forecast also predicted a depression to the south of Ireland which was not there in reality.

The Leutbecher *et al.* study addressed the question of whether this forecast could have been improved if there had been better observations of the

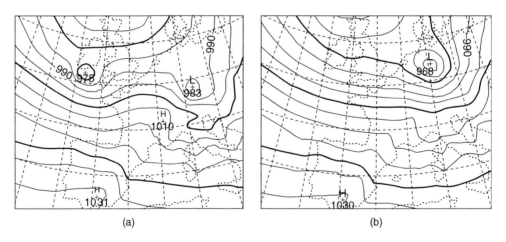

(a)   (b)

Figure 2.1   Mean sea level pressure (MSLP) fields at 12:00 UTC on 26 December 1999 from (a) a T+48 hour forecast from the ECMWF operational model and (b) the verifying analysis, that is 'truth'. (Reproduced by permission of ECMWF.)

developing system. Because the system formed and developed over the Atlantic Ocean, *in situ* observations were very sparse, largely limited to a few surface reports from shipping. The study used sophisticated mathematical techniques which followed the development of the system backwards through time. This allowed the 'optimal zones for observing' (OZOs) to be identified. These are essentially the parts of the atmosphere where small changes in the state of the atmosphere would have resulted in large changes to the development of the system. Put another way, these were the parts of the atmosphere in which the development of the Lothar storm was sensitive to small perturbations. It follows then that additional observations in these zones would have led to a better forecast. Leutbecher *et al.* tested this hypothesis by incorporating *synthetic observations* of wind and temperature similar to data from radiosondes in these zones. The synthetic observations were themselves created by taking vertical profiles from a numerical model forecast that simulated the Lothar storm more realistically than the ECMWF forecast shown in Figure 2.1. A total of 40 synthetic radiosonde observations were used, concentrated in the regions which had been determined as the optimal zones for observations. The new forecast of mean sea level pressure at 12:00 UTC on the 26 December 1999 is shown in Figure 2.2.

This forecast is a clear improvement on the original, with the low centre now in a better position and only 4 hPa too shallow instead of the 15 hPa

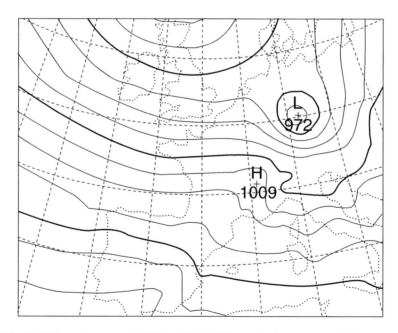

Figure 2.2   T+48 hour forecast of MSLP at 12:00 UTC on 26 December 1999 from the ECMWF hindcast run that included synthetic observations at the initial time. (Reproduced by permission of ECMWF.)

error in the original forecast. The spurious low to the south of Ireland is now no longer present.

This study showed that, had extra observations been available in the critical region where the storm was developing, the forecast could have been improved. Of course, it is not practical to position many new radiosonde launch sites over a limited oceanic region in the hope that they will lead to the occasional improvement of forecasts of potentially damaging storms. In future, though, it may be possible to deploy *targeted observations* in such critical regions using automated systems such as pilotless drones, dropsondes launched from neutral buoyancy balloons which are permanently floating in the upper troposphere or rocketsondes launched from moored buoys stationed at various fixed points across the oceans. In cases when a potentially damaging weather system is identified in an initial set of numerical weather forecasts, the critical regions for observation could be identified using a version of the techniques described by Leutbecher *et al.* Automated systems could then be activated to rapidly deploy the observing systems. The new observations could then be fed into the initial conditions of a new run of the numerical model. The viability of such targeted observation systems is an active area of current research. A field campaign in 2007 called the Greenland Flow Distortion Experiment (GFDex), for instance, showed that targeted dropsonde observations from aircraft flights improved the skill of forecasts of low pressure systems developing in the North Atlantic, although the improvements were small compared to the overall forecast error (Renfrew *et al.*, 2008).

It is an unfortunate fact that many of the potentially most destructive and difficult to forecast weather systems, such as the Lothar storm, form and develop in regions of the globe which are rather sparsely observed. Rapidly deepening mid-latitude cyclones tend to occur over oceanic regions and tropical cyclones develop exclusively over oceans too. The mid-latitude cyclone case is made worse by the fact that such systems often develop in regions with large amounts of upper cloud (e.g. near to jet streams), which limits the ability of satellites to provide useful vertical temperature and humidity profiles. This is another argument for the implementation of targeted *in situ* observations, which could provide more detail in the region of a developing storm than is currently available through remote sensing methods, such as satellite observations.

Whilst the Leutbecher *et al.* study showed that more observations in critical regions could improve the quality of forecasts of a rapidly developing depression, it was also the case that some operational NWP forecasts of that particular storm were more accurate than the original ECMWF forecast, *even without the benefit of extra observations*. The 40 synthetic radiosonde observations used to correct the ECMWF forecast were taken from the analysis of the MeteoFrance NWP model run at the same time. This had the same observational data available to it but managed to predict the storm much better than the ECMWF forecast. Just as interesting were the forecasts from the ECMWF ensemble. At the time ECMWF ran a 51-member ensemble, with each member differing only in its initial conditions. The sizes of the perturbations to the initial conditions were calculated to be of the same order of magnitude as the known errors in the observations, so that each ensemble member had an equally likely set of initial conditions. Of the 51 ensemble members initiated at 12:00 UTC on 24 December 1999, 17 predicted a depression over Europe at 06:00 UTC on 26 December which was *at least as deep* as the observed system (although a number of these depressions were in the wrong place). From this we can conclude that although observations are extremely important for initialising numerical weather forecasts, the quality of those forecasts is still sensitive to perturbations in the initial conditions which are of the same order of magnitude as the error bars on the available observations. And the fact remains that, a priori, there is no way of knowing which if any of a set of ensemble forecast members will turn out to be most accurate. However, if a significant proportion of an ensemble of forecasts predicts a certain event (in the Lothar case one third of the ensemble members predicted an intense storm) then this information can be used as guidance in the production of forecasts, for instance in the generation of a probability forecast of a particular event. How this might be done is discussed in detail in Chapter 5.

So a conclusion to this section on the importance of observations in weather forecasting would be that high quality observations are indeed crucial to the production of high quality weather forecasts. However, we also need to look at ways to address and account for observational errors and the inherent unpredictability of the atmosphere, which is still present even if our observations were considerably more comprehensive than they are ever likely to be.

2.3   An overview of the operational forecast process

The previous discussion has illustrated the nature of the operational weather forecasting problem. Clearly, careful analysis of the initial conditions at the start of the forecast period is crucial to ensuring a good forecast. This has been the case ever since operational forecasting started in the nineteenth century when the skill of meteorological analysis was developed and refined. Moving into the NWP age, the numerical methods used to produce the forecasts demanded an even more precise method of quantifying the initial state of the atmosphere and led to the incorporation of data assimilation into the forecast process, the details of which are examined in Chapter 4.

Since the determination of the initial conditions is so important in weather forecasting, a prior stage of the forecast process must be the timely making and collection of meteorological observations. This process is described in much more detail in Chapter 3. The more good quality observations are available, *as soon as possible after they have been made*, the better will be the specification of the initial conditions for the forecast. The timeliness aspect of the availability of observations cannot be overstated. Operational weather forecasting was only made possible following the development of the electric telegraph, which allowed weather observations to be transmitted almost instantaneously to a central forecasting office. In the modern era of electronic communications the rapid collection of weather observations for use in NWP is still a challenge. All operational forecasting centres apply a *data cut-off time*. This is the specified time after which no new observational data received at the forecasting centre will be included in the data assimilation cycle of a model run, because the forecasts need to be produced by a certain time in order to be useful. Because of the massively increased volume of meteorological observations available today, some data that may be useful in setting the initial conditions for a forecast will inevitably still miss this cut-off time despite the huge increase in the bandwidth available for their transmission.

Once the initial conditions have been determined, some sort of model which predicts how the state of the atmosphere will change with time needs to be applied in order to produce a forecast. In pre-NWP days this might have been a *conceptual model* such as the Norwegian frontal cyclone model, which describes the typical evolution of a particular type of weather system, or an *empirical model* (Box 2.2), which relates the current conditions to some future state through a statistical analysis of many years of past data. Other techniques used included simple advection methods, where the rainfall distribution, for instance, was forecast by assuming that the current distribution would simply move with the wind at a certain level of the atmosphere. *Analogue methods* were also widely used in which the current conditions were compared to previous similar situations and the forecast was based on the assumption that the atmosphere would go through a similar evolution. Clearly, given the previous discussion on how rather small changes to initial conditions can result in large differences in the evolution of the atmosphere, analogue methods were of very limited use, particularly at longer forecast lead times.

Box 2.2 Empirical forecast models

Prior to the advent of numerical weather prediction, meteorologists devoted a lot of time and effort to developing forecasting techniques based on statistical analysis of the behaviour of the weather in the past records. Many techniques were developed and published relating an individual weather parameter, such as the daytime maximum temperature, to some precursor variables, such as the thickness of the atmospheric layer between 1000 hPa and 850 hPa, taken from a representative overnight radiosonde sounding, and the cloud cover.

One of the simplest empirical techniques used by forecasters in the United Kingdom was the Mackenzie night minimum temperature technique. Given only the daytime maximum temperature (T_{max}) and the dew point temperature (T_d) at the time of T_{max} at the station, this method provided a forecast of the minimum temperature for the coming night, given by the formula:

$$T_{min} = 0.5(T_{max} + T_d) - K$$

where K is an adjustment factor calculated from a forecast of overnight wind speed and average cloud amount. Every station in the United Kingdom had its own table for the K factor based on analysis of the local conditions.

Such techniques allowed quick and simple methods for prediction based on a small amount of input data. Empirical forecast techniques were still appearing in weather forecasting manuals in the late twentieth century. However, they are fraught with problems and clearly a forecasting method based on the average behaviour of the weather over a long period will only produce forecasts which themselves appear skilful if averaged over a long period. In individual cases such techniques can be very wrong indeed. A large part of the skill of an operational forecaster using empirical methods was in recognising the situations where the method would not work and adjusting the forecast accordingly. For instance, with the Mackenzie minimum temperature technique, a change of air mass due to the passage of a front between the daytime maximum temperature and the coming night would invalidate the method. This technique also depends on an accurate forecast of overnight cloud cover, which itself is a very difficult variable to predict. Empirical methods to predict cloud amounts have been developed but are extremely unreliable. Cloud forecasts were more likely to be based on advection of the current cloud distribution, but prior to the availability of detailed satellite imagery this, too, was subject to a large degree of uncertainty.

Another problem with these techniques was that they were very often location specific. An empirical method for predicting, say, the occurrence of snow based on statistical analysis of weather data from a coastal station is unlikely to produce a method that can be applied equally well at an inland station at a considerable elevation above sea level.

At a modern NWP centre such as a national meteorological agency, the model used to make a forecast will be a full numerical model incorporating all the relevant physical processes that can determine the evolution of the atmosphere. Such models are highly complex and have only become possible as *operational* tools since the advent of high performance computing. Even though such models have been used routinely since the 1960s or 1970s, a combination of various factors has meant that it has only really been since the 1980s or even 1990s that they have become the dominant tool in forecast production. One factor is obviously the availability of computing power, which in the past meant that the physical complexity of the atmosphere had to be considerably simplified in the models, meaning that the forecasts themselves were often unreliable. Another factor was the difficulty of disseminating complex model output to forecasters at locations remote from the main NWP centre. During the 1980s, for instance, a UK Met Office weather forecaster at an airfield outstation only got to see a tiny fraction of the output from the UK Met Office NWP forecast. This was transmitted by facsimile machine and was often several hours old by the time it was received. Because of this, forecasters still relied on some of the rather out-of-date techniques listed above for many aspects of the forecast.

Once a model has produced a prediction of how the weather will evolve, this is not the end of the forecast process. At this point, the role of a human forecaster becomes crucial. Usually a small team of senior forecasters in a forecasting organisation will be involved in assessing the NWP forecast, deciding where the major uncertainties lie and communicating their conclusions to other forecasters involved in the production of forecasts for use by their particular customers. The fact that an NWP forecast takes several hours to produce is taken advantage of in this process. By the time the output from the model is available to the human forecasters, perhaps 2–4 hours of the forecast period have elapsed in real time. This means that the first few hours of the NWP forecast can be compared with actual observations to ascertain whether the forecast is on track during this initial period. Clearly, if there are large discrepancies between the model and reality in this short period it is unlikely that the forecast through the rest of the period will be particularly reliable, so the senior forecasters will look for other sources of guidance – perhaps forecasts from other NWP centres or even forecasts issued earlier. In most cases, however, differences between the forecast and reality in this early stage of the forecast are likely to be small and so can be taken into account when producing forecasts for customers. The senior forecast team will also be familiar with the characteristics of their own NWP model, so will know its systematic errors and biases. For instance, the model may be known to underpredict the deepening of Atlantic depressions in particular circumstances, so the forecasters can use this knowledge to make adjustments to the raw output from the model. The role of the senior forecasters in assessing and adjusting

NWP output, and then communicating their conclusions to other forecasters in the production process is discussed in Chapter 6.

Following their assessment of each new forecast as it appears, the senior forecasting team will then issue guidance to other forecasters whose primary task is to produce forecast products for customers. This guidance may be purely verbal or may include visual information as well, such as model forecasts that have been modified in some way by the forecasting team (for instance to enhance the rainfall rates in showers predicted by the model) or a set of forecasts from an ensemble in order to give a picture of the uncertainty in the forecasts. Guidance on uncertainty forms a key component of the issued forecast guidance, informing the forecast production teams of the levels of confidence they should attach to their forecasts and giving an indication of where the major uncertainties lie during the forecast period. In a period of rapidly changing, mobile weather systems driven by a strong jet-stream the typical uncertainties may be associated with the timing of the passage of individual fronts or depressions, or the intensity of these particular systems. In a more settled period, dominated perhaps by high pressure and a slowly evolving large scale pattern, the forecast uncertainty may well be associated with local scale detail, such as the presence or absence of low clouds overnight having an impact on the radiation balance and, hence, on overnight temperatures.

Individual forecasters involved in forecast production for customers then issue their forecasts on the basis of the NWP output, guidance from senior forecasters, their own specific knowledge of the local conditions in their forecast area and the exact requirements of their customers. Even in the NWP age, local knowledge is still an important component of the forecast process, as most NWP models are still not capable of representing all the small scale weather variability that may be crucially important to particular customers – for instance the persistence of fog at an airfield. A good knowledge of a customer's requirements is also very important and one reason why the forecast process still involves a human element despite the increasing sophistication of the NWP process. By having a really good knowledge of the weather elements that a customer regards as important, a forecaster can tailor their forecast to suit the customer. For instance, a forecaster working with military helicopter crews involved in battlefield training exercises will have a very different set of forecast criteria than one working with a supermarket chain using forecast information to predict demand of particular products in its shops. And both of these will be doing a very different job to a forecaster working with the electricity supply industry in order to produce hour-by-hour predictions of power output and demand across a country.

Once forecasts have been issued it would be wrong to think that forecasters then sit back and wait for the next run of the NWP model. In between forecasts, the job of monitoring the weather as it develops is also a key component of the forecasting process. It may be the case that the next run of the NWP model is 6 or 12 hours away, so if the weather starts to evolve in a way that differs from the previous forecast then it is not sufficient to simply sit and wait for new information from the computer model. Many forecast customers will require updates if the weather starts to diverge from the previously issued forecast, and some will have a specific set of criteria which will activate the need for an amended forecast to be issued. Hence, it is very important that forecasting teams monitor the weather continuously through the forecast period. Indeed, for some customers it may be the very short range detailed information, issued between the actual runs of an NWP model, that makes the biggest difference to their operations.

This continual monitoring and updating of the forecast is generally known as *nowcasting* – a somewhat jargonistic term but one which succinctly summarizes the process rather well (Box 2.3). The tools that forecasters use to do this are typically rapidly updating observational systems which provide quite a lot of local detail over the area of interest. Meteorological radar systems, for instance, can provide updates of the location and intensity of precipitation as frequently as every five minutes, giving forecasters a very detailed picture of how the prediction of the precipitation pattern is comparing with observations. In regions prone to heavy thunderstorms and even tornado outbreaks the meteorological radar network is often the most crucial tool in the forecaster's toolkit. NWP models can predict general areas which may

---

Box 2.3 Some sporting examples of nowcasting

Several major sporting events and sports teams are prepared to pay considerable sums of money to forecasting companies or organisations and sometimes employ their own meteorologists in order to access nowcasting advice. Lightning is a potentially life threatening hazard to golfers and so some major tournaments will take a nowcasting service from a forecast provider in order to ensure the safety of the players whilst minimising disruption to play. Major outdoor tennis tournaments also take nowcasting services in order to get advance warning of when the courts may need to be covered due to the threat of rain. Some Formula 1 motor racing teams employ a meteorologist in their pit crew, advising the drivers in advance of the need to change tyres – advice which may give them a few vital seconds advantage over rivals. Small changes in wind can also have big impacts on the outcome of a sailing race, so many Olympic sailing teams employ their own meteorologists who study the typical weather patterns of the race venues and are on hand to give advice on what to expect during the course of each race.

be prone to this type of severe weather several days ahead, but it is only by continually monitoring the observations that forecasters are able to issue actual warnings of where a tornado may actually touch down, giving people sufficient time to get to shelter.

On a larger scale, satellite imagery is also an important nowcasting tool, with the current generation of geostationary satellites providing a new image every 15 or 30 minutes. These images allow forecasters to watch the regional scale evolution of the atmosphere. An experienced forecaster will be able to use the imagery to pinpoint the location of features such as depressions, troughs, fronts, frontal waves and jet streams, which can then be compared with predictions from NWP models in order to assess the veracity of the forecast. Crucially, satellite imagery also allows forecasters to observe how such features are changing in intensity. For instance, a rapidly growing area of high cold cloud that appears to be getting larger and colder on an infrared satellite image may well be showing a region of atmospheric ascent where cloud is forming and deepening rapidly. Such regions are often associated with developing waves on frontal systems, which is something that the current generation of NWP models sometimes has difficulty predicting, even at short lead times. The appearance of an emerging 'cloud head' (Carlson, 1980) on satellite imagery in association with the signs of a frontal wave can indicate that the wave is developing into a cyclone with its associated low level wind circulation.

One example of how careful monitoring of satellite imagery led to a successful modification of a forecast comes from the same Lothar Storm of the 26 December 1999 discussed in Section 2.2. The UK Met Office forecast originally followed a similar evolution to the ECMWF forecast, which failed to predict the developing storm. However, a senior forecaster at the UK Met Office studying the satellite imagery noted signs of a rapidly developing frontal wave in the centre of the Atlantic. Checking against ship reports of surface pressure revealed that there were observations of rapidly falling pressure but these were being rejected by the NWP model data assimilation as being too far from the model background field to be reliable (Chapter 4 provides a fuller description of the data assimilation process). Based on the evidence from the ships and the imagery together, the forecaster was able to force the model to take more account of the ship observations and the result was a forecast that was much closer to capturing the reality of the Lothar storm than that which would have occurred without the intervention (Figure 2.3).

The constant monitoring and updating of the forecast throughout the forecast period is really an acknowledgement of the unpredictability of the atmosphere. It is clear that however advanced NWP methods and models become, there will still be times when the forecast diverges from reality at rather short forecast ranges. Hence the job of nowcasting is necessary so

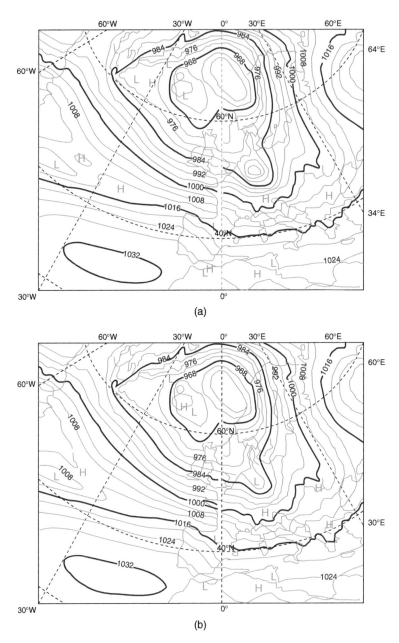

Figure 2.3 Forecasts of the mean sea level pressure distribution associated with the Lothar storm of 26 December 1999 from the UK Met Office NWP model. (a) shows the storm as predicted after forecaster intervention whereas (b) shows the forecast which the model would have produced with no intervention. See Figure 2.1b for the verifying analysis of this storm. (© Crown Copyright 2000, Met Office.)

that adjustments can be made to the forecast in the light of the developing conditions. In Chapter 5 a detailed look is taken at the use of ensemble forecasts, which are designed to give some idea of the uncertainty of each particular forecast and communicate that uncertainty through the use of probability forecasts.

Summary

- Deterministic forecasting is limited by the non-linear ('chaotic') nature of the atmosphere.
- The deterministic limit for weather forecasting is about 14 days but in some cases it may be considerably shorter than this.
- Accurate specification of the initial conditions for a weather forecast is crucial for determining how the forecast will evolve.
- The weather forecasting process involves:
  1. Collection of observations
  2. Using those observations to specify the initial conditions for the forecast
  3. The use of a model to extrapolate the state of the atmosphere forwards in time
  4. Assessment of the output of that model by experienced forecasters
  5. Production of forecasts for customers.

3

Meteorological Observations

3.1 What do we need from a meteorological observing system?

Observational systems and computer models, and their respective operators and developers, often vie for scarce financial resources. The truth is, however, that each needs the other and it is an aim of this chapter and book to demonstrate why this remains so and the tremendous synergy which can result. The availability of upwind measurements makes it possible, of course, to make one of the easiest forms of weather forecast, namely the simple downwind advection of weather conditions, ignoring possible development or dissipation of those conditions. There is a skill, of course, in identifying when and where such an 'advection forecast' is most likely to succeed. We all make short range and simple advection forecasts, sometimes known as *nowcasts*, without the use of data or technology by simply looking at the state of the upwind sky!

It is first worth recalling the many good reasons for making weather observations, including some which are not immediately relevant to the main subject of this text, weather forecasting. Some of the earliest observations using scientific instrumentation were made in the late sixteenth and early seventeenth centuries, underpinned by the pioneering work of the inventors of the thermometer and barometer. Of course, visual observations and descriptions of the weather long pre-dated these instrumental records but, very significantly, access today to the longest quantitative high-resolution measurements on which we heavily depend for the identification of trends in recent climate must acknowledge the importance of the work of these scientists and historians. Many of the very longest measurement records derive from Europe. Balling *et al.* (1998) report, for example, on station-derived European temperature trends based on records commencing in 1751. The very oldest series, such as the UK Central England Temperature (CET) monthly time series (Manley, 1974), begins as far back as 1659. The relevance of those earliest instrumental records is not lost either on present

Operational Weather Forecasting, First Edition. Peter Inness and Steve Dorling.
© 2013 John Wiley & Sons, Ltd. Published 2013 by John Wiley & Sons, Ltd.

day numerical weather prediction (NWP), since their temporal and spatial variations triggered much curiosity and scientific endeavour with respect to improved understanding of the causes of such variations, and of atmospheric behaviour in general. The sum of this research, over centuries, underpins the physics included in today's NWP models. The aviation era, particularly World War II, triggered new questions about atmospheric behaviour for meteorologists to grapple with, not least the science behind *jet streams* and their relevance to wartime missions. It is not surprising then that some National Meteorological Services have, or have had, strong links with the military. During the early part of the twentieth century the UK Met Office was part of the Air Ministry and the Saudi and Italian Met Services have strong links to their respective air forces. Air bases and civilian airports continue to be convenient, secure and necessary places for meteorological measurements, one reason for an apparent mismatch between surface weather data availability and population centres.

Measurements of extremes, especially those forming part of long-term records, continue to be used in forming the basis for the design of major infrastructure, such as bridges and dams, facilitating the assessment of the all-important *return periods*. New measurement records naturally attract media interest too, be it the strongest wind gust ever recorded on Earth (408 km/h during Tropical Cyclone *Olivia* on 10 April 1996 at Barrow Island, Australia), a new national maximum twenty-four hour rainfall total (such as in Cumbria, UK, in November 2009; Eden and Burt, 2010) or a most active tropical cyclone season in an ocean basin. Claims of new official records are investigated in depth before they can be confirmed and adopted.

Regular observations also underpin, of course, the calculation of long-term averages or *climate normals* as they are known. It is area-specific *climate normals* which we have in mind when we pack our holiday suitcases. Such averages are usually calculated over a 30-year period, although in a rapidly changing climate such an averaging period can be misleading and, in addition, be at odds with our most recent memory. These averages, or *climatology*, are, nevertheless, very valuable benchmarks against which to measure the performance of NWP models in verification procedures (Chapter 8). If our models cannot out-perform *climatology* then they clearly do not demonstrate forecast *skill*.

The era of NWP has introduced additional demands on the observational system, in some cases creating a tension with other users of observational data. NWP prioritisation with respect to observations tends to focus on requirements for data assimilation, as discussed in Chapter 4 (including measurements made over otherwise data sparse regions), and for model verification, with perhaps less weight given to the value of long-term records at particular locations. The latter, of course, are of especial interest, meanwhile, to climate scientists and engineers.

The modern forecasting era has emphasised the terrific and growing contribution of remotely sensed data in providing continuous and global coverage, starting with the first satellite image from TIROS in 1960 (Mohr, 2010). Again, these observations have been instrumental in improving our scientific understanding and awareness, for example in providing a more accurate record of tropical cyclone records over the previously poorly monitored remote oceans.

The operational planning and sharing, internationally, of meteorological observations through the Global Observing System and Global Telecommunications System (GTS) is a triumph of collaboration (Branski, 2010) on which the assimilation and verification schemes within NWP models depend, interrupted only in times of conflict between countries in dispute with one another.

Observational data have a myriad of other applications, for example defining when it is safe to carry out particular tasks or pursuits (operating a crane, sailing a boat) or when an insurance policy should pay out (in the event of property damage during severe weather, loss of attendance revenue due to poor weather during a community event or the triggering of a weather derivatives arrangement) and in providing supporting evidence of relevance to a legal case (perhaps relating to icy roads or burglaries through open windows). Satellite monitoring is a powerful way of regularly checking on the evolving health of our environment in general and in supporting disaster recovery.

3.2 Data transmission and processing

It might seem obvious but one of the most valuable uses of meteorological observations is in *checking* that the weather forecast, issued some hours previously, is still supported by the latest real-world measurements; the confidence placed by users in a forecast is quickly dented when the view out of the window strongly contradicts the words of the radio or TV weather presenter! We will see, in Chapter 4, that the measurements are also critical in the development of an accurate NWP model analysis, an essential initial ingredient for any accurate forecast. It follows that, for forecasters and supporting systems to be able to continuously undertake such checking, a number of things are crucial, especially bearing in mind the global scale of the measurements:

1. State-of-the-art communication systems are needed so that the measurements can be shared and accessed as quickly as possible – observations are particularly perishable in application to forecasting. The meteorological community relies upon the GTS for the efficient and effective dissemination

of data and 'comms' are also crucial, of course, in the rapid dissemination of forecasts to users (Branski, 2010).

2. The format in which data are transmitted should adhere to internationally agreed standards; this ensures that the data are presented in unambiguous and concise ways. Coding systems are common in meteorology and metadata (information about the data) has to be comprehensively documented.

3. Quality control systems should be an integral part of the weather forecast system and should filter and/or flag unreliable measurement data. A discussion of how unreliable data might be recognised is given in Chapter 4 (Section 4.3.3).

4. Software visualisation systems should present the measurement data in a time series or spatial form that is easy to interpret and to compare with forecasts. The multidimensional nature of atmospheric conditions demands ever more powerful software solutions for generating the required presentation.

5. International collaboration which continuously reviews the instrument hardware which is approved for achieving the measurement standard requirements of the community. Much of this work is undertaken under the auspices of the World Meteorological Organization (WMO) as part of the Global Observing System (Nash *et al.*, 2010). A description of the role of the WMO is given in Box 3.1.

Box 3.1 The role of the World Meteorological Organization (WMO) in observing systems

A major function of the World Meteorological Organization (WMO), under its convention, is 'To facilitate worldwide cooperation in the establishment of networks of stations for the making of meteorological observations as well as hydrological and other geophysical observations related to meteorology, and to promote the establishment and maintenance of centres charged with the provision of meteorological and related services'.

WMO members coordinate measurement networks in space, in the atmosphere, on land and over the oceans. Integration of the WMO Global Observing System (GOS) with WMO co-sponsored observing systems such as the Global Ocean Observing System (GOOS), Global Terrestrial Observing System (GTOS) and Global Climate Observing System (GCOS) is underway, leading to the WMO Integrated Global Observing Systems (WIGOS). 'WMO monitoring and observing systems are a core component of the Global Earth Observation System of Systems (GEOSS), aimed at developing a comprehensive, coordinated and sustained Earth observation system of systems'.

Meanwhile the WMO's Instruments and Methods of Observation Programme (IMOP) is concerned with ensuring that meteorological instruments of all forms are fit for purpose and standardised to facilitate inter-comparison.

http://www.wmo.int/pages/themes/observations/index_en.html

3.3   Observing platforms

It is important to recognise that weather forecasting in the early twenty-first century requires supporting observational systems which go far beyond the original focus, over one hundred years ago, on manually collected and recorded *in situ* measurements of the atmosphere. The remarkable first step, in 1960, into the remote sensing era (the launch of the TIROS satellite, Figure 3.1), has been followed, equally dramatically, by monitoring infrastructure which underpins our present day appreciation that, to improve the accuracy of weather forecasts, it is necessary to also monitor, in real time, the oceans, the cryosphere and the atmospheric chemical composition

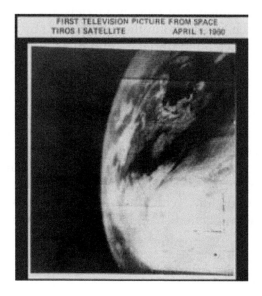

Figure 3.1   TIROS-I First Weather Satellite image, 1 April 1960. The picture shows the New England Coast of the United States of America and Canada's Maritime Provinces, north of the St. Lawrence River. (Reproduced from Mohr (2010), with permission from WMO Geneva.)

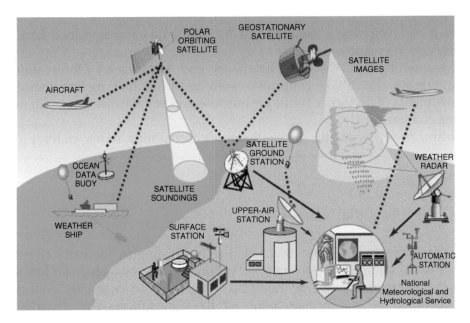

Figure 3.2 Global Observing System. NMHS: National Meteorological and Hydrological Service. (Reproduced from Branski (2010), with permission from WMO, Geneva.)

(so-called *Chemical Weather*). It is only by monitoring this broader environment that we can more accurately simulate, in models, the way in which energy is partitioned and exchanged within the whole Earth System (Figure 3.2). In addition, progress in scientific understanding has also led to an increased requirement for physical and chemical measurements in the stratosphere (10–50 km above the Earth's surface) as we better appreciate the relevance of stratosphere–troposphere interactions for realistic model simulations (more details are given in Section 7.1.3). Figure 3.3 shows the broad range of both *in situ* and remotely sensed measurements which currently underpin the data assimilation system of a state-of-the-art forecasting centre. Our real-time need for cost effective information about the remote environment has required the development of monitoring systems which are low power, low maintenance and, increasingly, automatic, with obvious implications for technological development and the associated changing staff skill requirements.

3.3.1   In situ *networks*

Before the era of remote sensing, meteorological measurements were all made *in situ*. Coding systems were developed for the concise recording

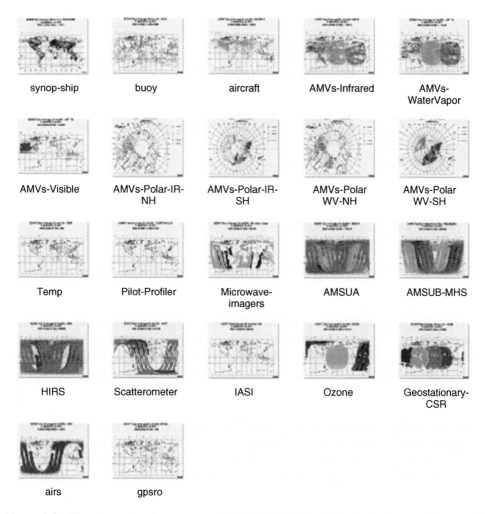

Figure 3.3 The observational sources used by the ECMWF Data Assimilation System. The network is dominated by satellite-based systems. (Reproduced by permission of ECMWF.)

and efficient exchange of surface measurements made over land (SYNOP; METAR), on board SHIP (SHIP), by weather buoys (BUOY) and, latterly, by aircraft (AIRCRAFT) and for upper-air data collected by weather balloons with and without radiosondes attached (TEMP; PILOT). While it is not necessary today for forecasters and other users to recall *all* the details of these coding systems (we rely on computers for that), the data are still recorded and transmitted in this coded form, for reasons of speed of transmission and the avoidance of any ambiguity.

A major change which has occurred in the last 20–30 years has been the rapid transition from manual measurements made by skilled observers

  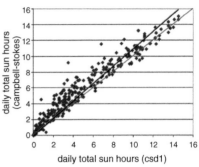

Figure 3.4 Campbell-Stokes Sunshine Recorder (left), CSD1 Electronic Solar Radiation Sensor (centre) and the relationship between respective co-located measurements made at Camborne, Cornwall, UK (right). (© Crown Copyright 2003, Met Office.)

to the dominance of automatic recording systems. The transition is not without compromise, however, since cost savings and the removal of human bias/error are partially offset by some technological limitations, such as in the reporting of cloud type, and disturbance to long records which can occur as a result of the introduction of a new sensor. Great care is needed in dealing with the latter and a good example is the replacement, in the 1990s, of the Campbell–Stokes Sunshine Recorder by the CSD1 electronic sensor (Figure 3.4), with both devices operated in parallel for a lengthy period in order to establish a reliable conversion factor. The technical advantages of moving to an automated sensor were significant: (i) the sunshine recorder, operating on the principle of a glass sphere burning a hole on a chart during periods of sufficient sunshine intensity, was known to overestimate sunshine on days of patchy cloud, when the temporal response of the device was insufficient to accurately reflect the many changes in sunshine conditions; (ii) the electronic sensor was better suited to providing continuous estimates of solar radiation flux, of use in model data assimilation systems and in model output verification, as well as the 'sunshine hours' more familiarly seen in the media and used in climatology.

Let us compare and contrast the different *in situ* measurement networks which are now used, every day, in the data assimilation systems of the world's global NWP models. Figure 3.5 shows the relevant *in situ* networks which formed the basis of observational assimilation in the 00:00 UTC run of the ECMWF global model on 17 April 2012. Five things are immediately clear:

1. the profusion of land-based surface measurements (SYNOP) in Figure 3.5a relative to those over the sea (SHIP and BUOY) shown in Figures 3.5a and 3.5b, despite the Earth being 70% ocean covered;

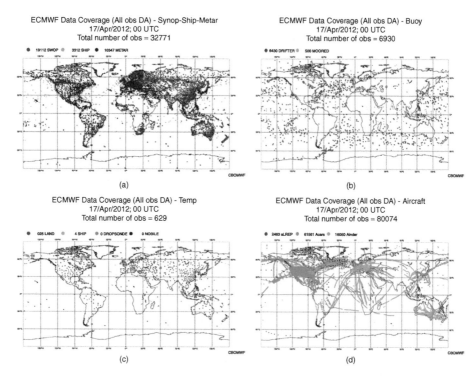

Figure 3.5 *In situ* (a) SYNOP/SHIP, (b) BUOY, (c) TEMP and (d) AIRCRAFT networks used by the 17 April 2012 00:00 UTC run of the ECMWF Global Model data assimilation system. (Reproduced by permission of ECMWF.)

2. the very uneven spread of land measurements between continents in Figure 3.5a;
3. the orders-of-magnitude lower number of radiosonde balloon launching stations in Figure 3.5c relative to regular surface measurement stations;
4. the influence of the major commercial airline routes on the spatially varying density of aircraft-based measurements to support operational meteorology (Figure 3.5d);
5. that *in situ* measurements remain a vital part of the data assimilation system, complementing (and helping to calibrate) the burgeoning remote sensing systems. The cost of maintaining measurement networks is significant and this calls for both regular 'fit for purpose' reviews and much operational quality control.

SYNOP/on board ship (SHIP)/METAR

Figure 3.6 shows an example of all the ground-based SYNOP measurements made at a particular time, 06:00 UTC on 15 February 2010, over the British Isles (data from SYNOP stations are available at least three hourly, and often hourly, across the GTS network). This map provides an incredibly concise and comprehensive picture of the surface conditions experienced at this time across the country, dealing with all the surface weather variables which one might be interested in such as sea-level pressure and pressure tendency, air temperature and dew point temperature, wind speed and direction, cloud amount, height and type, visibility and precipitation type. Without the coding system we would find it necessary to present all this information on several separate maps, thereby making it harder for the human brain to establish all the interesting and important linkages between the variables. The measurements shown in Figure 3.6 make a small but important contribution to the computer generated construction of the model analysis (described in the data assimilation section of Chapter 4) presented in Figure 3.7, with isobars and fronts added. To the uninitiated, the measurements are, however, displayed in Figure 3.6 in a format which is not immediately interpretable without a little knowledge of the SYNOP coding system. Figure 3.8 shows how the data are organised at each measurement location. (The full SYNOP code is available at http://weather.unisys.com/wxp/Appendices/Formats/SYNOP.html.) Raw real time and recent SYNOP reports are available at http://weather.cod.edu/digatmos/syn/.

The METAR code is reserved for regular measurements which are made at airfields, focusing on those parameters of greatest importance to aviation, for example cloud height and visibility. The full METAR code and current data are available, respectively, at http://www.nws.noaa.gov/oso/oso1/oso12/document/guide.shtml and http://www.aviationweather.gov/adds/metars/.

Other surface measurement networks over land do exist, of course, such as the roadside automatic weather stations (Figure 3.9) designed to provide alerts regarding dangerous driving conditions and the possible need for road gritting (Chapter 6), but these are mostly used in the specific task of providing and verifying road forecasting services. In addition, the ownership of such stations is more likely to be spread across many local organisations.

Dependency on observations from on board dedicated weather ships has significantly declined, again due to the attraction of automatic systems. Their usefulness over the ocean, upwind of major continents such as Europe, was however evident and helped make the case for investment in satellite technology to cover these otherwise data sparse regions. Today, observations from ships are provided under the Voluntary Observing Ship (VOS) Programme, consisting of approximately 4000 ships, around 1000 of which report each day.

Webcams (skycams), while not being a formal part of the observing system, can be helpful in providing imagery of sky conditions at remote locations, helping a forecaster to monitor developing conditions and test the usefulness of alternative automated instrumentation.

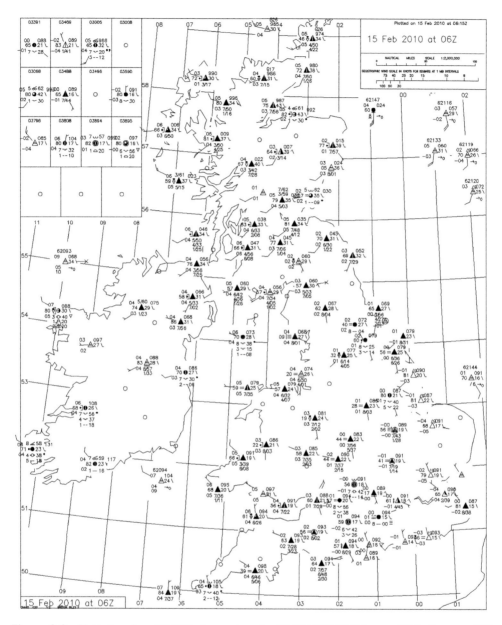

Figure 3.6 Plotted surface SYNOP observations for 06:00 UTC 15 February 2010. Courtesy UK Met Office. (© Crown Copyright 2010, Met Office.)

BUOYS

With so much of Earth's surface covered by ocean, which strongly influences the development and modification of our weather, having access to both drifting and near-shore moored weather buoy data provides a vital early

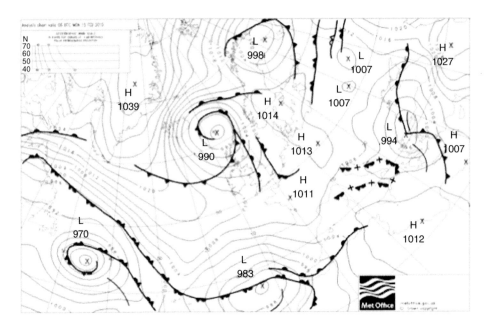

Figure 3.7   MSLP analysis for 06:00 UTC 15 February 2010. (© Crown Copyright 2010, Met Office.)

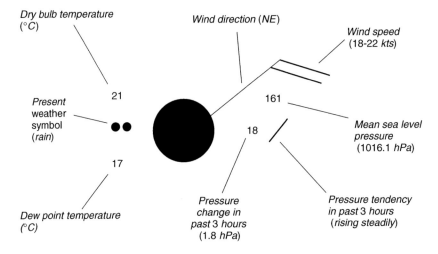

Figure 3.8   SYNOP plotting model showing the location of information on key meteorological variables around the 'station circle'

warning of the approaching conditions. Many countries have recognised the significant contribution which marine data can make, particularly during extreme weather conditions, for example in Australia (Schiller *et al.*, 2009).

The National Data Buoy Centre (NDBC), within the US National Weather Service, is an example of a coordinated buoy network consisting of 90 buoys

Figure 3.9 Roadside Automatic Weather Station, in inclement weather, at the entrance to the Mont Blanc Tunnel, Italy

measuring air temperature, barometric pressure, wind speed and direction, sea surface temperature and wave height and period around the US coast (Figure 3.10). Buoy mooring type depends on the water depth and buoy hull type – in coastal waters an all-chain mooring is common while in deeper water a combination of chain, nylon and polypropylene is generally used. In Europe, buoy data exchange is coordinated under the EUMETNET Composite Observing System. Meanwhile, the WMO and Intergovernmental Oceanographic Commission's data buoy cooperation panel (DBCP) also coordinates the use of autonomous data buoys in locations otherwise poorly sampled by *in situ* observations.

More information on ocean buoys used for measuring subsurface conditions, which have particular relevance to setting the initial conditions in ocean models, can be found in Section 7.2.

Figure 3.10   A moored meteorological buoy.
(Source: http://www.ndbc.noaa.gov/mooredbuoy.shtml.)

TEMP (Radiosonde) and PILOT (Rawinsonde)

Vertical *in situ* profiles in the troposphere and lower stratosphere of temperature, humidity, wind and pressure, typically made twice a day at 00:00 and 12:00 UTC, still make one of the most important contributions to improving the accuracy of model analyses. The current global network comprises approximately 1300 stations, of which maybe half will provide data to be used in any given NWP global model data assimilation cycle. Monitoring the movement of ascending PILOT balloons provides a vertical wind profile while the addition of a radiosonde, with integral sensors, to the balloon payload provides the temperature and humidity (TEMP) information (Figure 3.11). Many different tracking systems have been used over the years but the current ubiquity of GPS technology has made this task all the more easy, as has the availability of automatic balloon release systems (Figure 3.11) although manual release is still common in the developing world. These vertical profiles are invaluable in, for example, identifying stable and unstable atmospheric layers, in highlighting

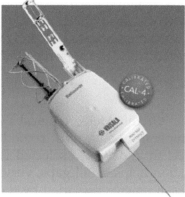

Figure 3.11   Ship-borne balloon launching system (left) and radiosonde (right). (Courtesy of Vaisala Oyj.)

regions of vertical wind shear, in accurately locating the freezing level and tropopause height and in providing 'truthing' measurements for comparison with satellite-derived data. Forecasters, and particularly those involved in forecasting for aviation, still use information from radiosonde soundings as part of their forecast production process. One issue with balloon-borne measurements is that several different sensor packages are used around the world, all of which have different instrument error characteristics that need to be dealt with, particularly when the measurements are used in NWP data assimilation.

The radiosonde soundings plotted as tephigrams in Figure 3.12 show the layer close to saturation below 800 mb at two stations ahead of the

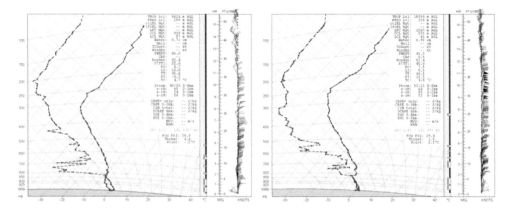

Figure 3.12   Tephigrams displaying the radiosonde soundings from Herstmonceux, southeast England (Station 03882 − left) and from Nottingham, central England (Station 03354 − right) at 00:00 UTC on 15 February 2010 for comparison with Figure 3.7 (global radiosonde data available from http://weather.uwyo.edu/upperair/sounding.html)

approaching surface warm front shown in Figure 3.7, with somewhat drier air above this level at Herstmonceux, further from the front. A small temperature inversion is shown at 820 mb on the Nottingham ascent as the balloon likely intersects the warm frontal boundary.

AIRCRAFT

The Aircraft Meteorological Data Relay (AMDAR) system provides wind, temperature and, most recently, humidity observations at aircraft cruising level in addition to selected levels during ascent and descent. This information helps to fill gaps, very cost effectively, in the network of upper-air measurements provided by the radiosonde network, and thereby supports NWP data assimilation systems. Of course, most of the information from instruments mounted on commercial airliners comes from very high levels in the troposphere so does not provide much vertical resolution. However, the levels at which commercial aircraft fly are often the levels which correspond to the level of jet streams which are crucial to determining the development and movement of weather systems right through the depth of the troposphere. The measurements are made primarily by commercial aircraft but also by some military and private operations; the data volume has grown significantly since the starting point in 1986. Figure 3.13 shows the daily number of AMDAR

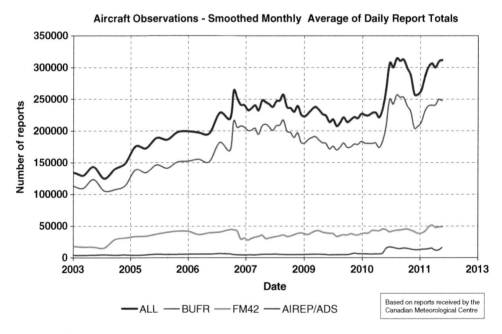

Figure 3.13 Volume of AMDAR reports over the period 2004–2009. BUFR, FM42 and AIREP refer to different data formats. (Source: WMO Geneva. Reproduced with permission.)

reports since 2004. All matters relating to the development of AMDAR, including quality control, training and network enhancement, are coordinated by the AMDAR panel (http://www.wmo.int/amdar/AMDARPanel.html). AMDAR systems are organised at the regional level and each requires a participating airline which has aircraft with communications, avionics and meteorological sensors that are compatible, ground or satellite links to the respective National Meteorological Service, connection to the GTS, data quality monitoring and data management capability. The benefits to participating airlines are clear, given the particular relevance of AMDAR data to nowcasting of thunderstorms, differentiation between precipitation types, wind shear and turbulence identification, low cloud positioning and fog formation (see also Section 6.2.3).

3.3.2   Remote sensing

Remote sensing generates incredible volumes of data which also demand efficient processing and archiving. Regarding satellites in particular, each sensor on board a satellite has an associated algorithm which has been developed for converting the raw satellite measurements into meaningful meteorological quantities. So when we see the increasingly familiar form of a satellite *image*, we must realise that this is the end product of a complex sequence of initial research, followed by operational data processing in which binary data are converted into a spatial representation of the variable of interest. For context, it is interesting to note that even before a satellite is launched, with its on-board measurement sensors, a decade or more of painstaking design, preparation and construction can be required before we can benefit from the operational data streams. The considerable expense of satellite meteorology leads to collaborations being forged between countries to jointly fund major monitoring initiatives.

Satellite remote sensing

Figure 3.14 shows the present day WMO Global Satellite Observing System, while Figure 3.15 shows the growth of satellite sensor data routinely assimilated into the ECMWF operational model. The system has developed remarkably since the first image provided by the polar orbiting US National Oceanic and Atmospheric Administration (NOAA) TIROS satellite in 1960 (Figure 3.1), a particularly significant element being the expansion from imaging to vertical soundings. Fundamentally, the system consists of geostationary weather satellite platforms (first launched in the 1970s) orbiting Earth at a height of 35 800 km and low-Earth-orbit (or polar orbiting) satellite

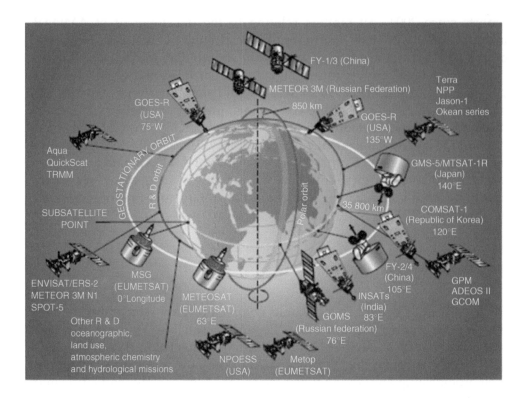

Figure 3.14   The WMO Global Satellite Observing System. (Reproduced from Mohr (2010), with permission from WMO Geneva.)

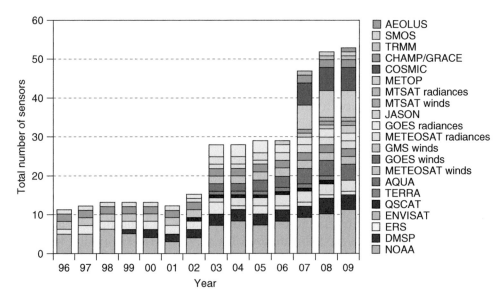

Figure 3.15   Number of satellite sensors used in ECMWF operational data assimilation. (Reproduced by permission of ECMWF.)

platforms at the much lower altitude of 850 km and with orbiting times of the order of 100 minutes. Equally fundamental are the four different types of sensing aboard these platforms, all of which involve radiation measurements, consisting of passive monitoring of surfaces, passive sounding of the atmosphere, active sensing of the surface and atmospheric sounding using GPS Radio Occultation. It is important to note that satellite instruments measure radiances at the satellite and not the meteorological variables which are the immediate requirements of NWP data assimilation systems (Section 4.3.6). As a result, one of two approaches is required to facilitate the data assimilation process. One is to generate synthetic observations from the model which are then compared with the actual measurements in order that the model state can then be nudged towards the observations. Or, alternatively, satellite measured radiances are processed using the radiative transfer equations, constrained using model output, for the so-called *retrieval* of wind, temperature or humidity which can then be assimilated into the model directly (Collard *et al.*, 2011).

Geostationary satellites have the advantage of, collectively, providing wide coverage at high temporal and spatial resolution, particularly useful for feature tracking, but the disadvantages of being at too high altitude to be able to use important regions of the electromagnetic spectrum, such as the microwave, and needing broad sensor channels due to relatively weak signals. The lower altitude polar orbiters meanwhile offer high spectral and spatial resolution but only moderate temporal resolution, meaning they are not so helpful for feature tracking and nowcasting.

Those satellites shown in Figure 3.14 with a stated longitude position, such as MSG (Meteosat Second Generation) at 0° longitude, form the geostationary network and these satellites remain in a fixed position above Earth by orbiting at the same speed as the Earth rotates. By way of example, MSG-2 currently provides continuous visible and infrared imaging, using the Spinning Enhanced Visible and Infrared Imager (SEVIRI) passive imaging radiometer, based on twelve spectral channels, on a 15-minute repeat cycle at 3 km resolution. Table 3.1 shows details relating to each of the radiometer channels, their respective sensitivities to radiances from gases at different atmospheric levels and their resulting individual purposes. The current GOES (Geostationary Operational Environmental Satellite) series of geostationary satellites, operated by the US National Environmental Satellite, Data and Information Service (NESDIS), also carry an 18-channel infrared passive *sounder* to acquire vertical temperature and humidity profiles. Geostationary satellites are also operated by China, India, Japan and Russia and standby satellites are placed in orbit in case of malfunction. One of the most important uses of geostationary satellite data for nowcasting and NWP is in the tracking of features, such as clouds, for the production of Atmospheric Motion Vectors (AMVs), although being certain of the atmospheric level

Table 3.1   SEVIRI Spectral Channel characteristics and applications (adapted from Schmetz et al., 2002)

Channel number	Channel name	Characteristics of spectral band (μm)			Main gaseous absorber or window	Purpose
		λ_{con}	λ_{max}	λ_{min}		
1	VIS0.6	0.635	0.56	0.71	Window	Cloud detection, cloud tracking, aerosol, land surface and vegetation monitoring
2	VIS0.8	0.81	0.74	0.88	Window	Cloud detection, cloud tracking, aerosol, land surface and vegetation monitoring
3	NIR1.6	1.64	1.50	1.78	Window	Discriminates between snow and cloud, and between ice and water cloud; aerosol information
4	IR3.9	3.90	3.48	4.36	Window	Low cloud and fog detection; nighttime land/sea temperature; winds from cloud tracking
5	WV6.2	6.25	5.35	7.15	Water vapour	Water vapour and winds
6	WV7.3	7.35	6.85	7.85	Water vapour	Water vapour and winds
7	IR8.7	8.70	8.30	9.10	Window	Quantifies high cirrus cloud; discriminates between ice and water clouds
8	IR9.7	9.66	9.38	9.94	Ozone	Ozone radiances reveal wind patterns in lower stratosphere
9	IR10.8	10.80	9.80	11.80	Window	Sea/land/cloud-top temperatures; cirrus cloud and volcanic ash
10	IR12.0	12.00	11.0	13.0	Window	Sea/land/cloud-top temperatures; cirrus cloud and volcanic ash
11	IR13.4	13.40	12.4	14.4	Carbon dioxide	CO_2 absorption – when cloud free, reveals temperature information in lower troposphere
12	HRV	Broadband		(about 0.4–1.1)		High resolution visible applications

being monitored can sometimes be difficult. Clear sky radiances are also very valuable in the construction of an NWP analysis (Section 4.3.6).

Today's polar orbiting satellites host advanced hyperspectral infrared sounders such as the Infrared Atmospheric Sounding Interferometer (IASI) on board the EUMETSAT MetOp satellite, operating at thousands of wavelengths and providing vertical temperature and humidity resolutions of the order of 1 km. Microwave sounders have also been introduced to overcome sensitivities of infrared sounders to cloud presence. Polar orbiting satellites are also the platform for sensors which infer ocean surface wind speeds from their effect on ocean surface wave characteristics, namely passive microwave radiometers (such as the Special Sensor Microwave Imager, SSM/I) and active radar microwave scatterometers, which work on the backscatter principle from ocean surface waves (such as the Advanced Scatterometer, ASCAT; Figure 3.16). NWP model winds are often useful for constraining the associated wind directions. Also on low-Earth-orbit platforms are receivers which sense the radio signals emitted by GPS satellites. These receivers

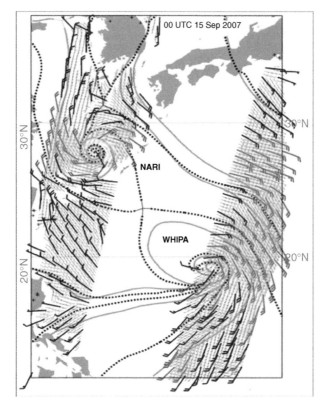

Figure 3.16   ASCAT wind barbs assimilated at ECMWF. Bold barbs show the data used in the assimilation and smaller pale barbs show the data prior to thinning. Contours show ECMWF surface wind streamlines (grey: first guess; dotted: analysis) for two typhoons, NARI (top left) and WIPHA (bottom right), to the south of Japan on 00:00 UTC 15 September 2007. (© EUMETSAT.)

are capable of quantifying the extent to which GPS signals are bent and delayed by the state of the atmosphere, especially the refractive index of the upper troposphere and lower stratosphere which is a function of pressure, temperature and humidity, providing relatively high vertical resolution of a few hundred metres.

We should briefly also discuss the significant role played by satellite remote sensing in monitoring sea surface temperatures (SSTs), sea ice and changing atmospheric composition, and the importance of these for NWP modelling. By combining satellite infrared and microwave imaging – using sensors aboard both geostationary and polar orbiting satellites – with *in situ* marine measurements, the UK Met Office's Operational Sea-surface Temperature and Sea Ice Analysis (OSTIA) system generates daily updates of a combined SST/Sea Ice concentration product at approximately 6 km resolution (Stark *et al.*, 2007; Donlon *et al.*, 2011). Figure 3.17 shows the SST

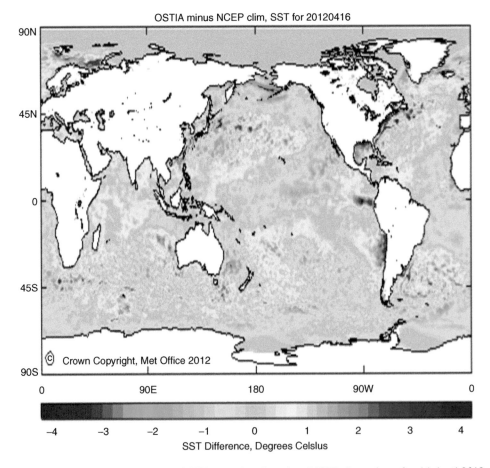

Figure 3.17   OSTIA derived global SST anomalies, based on NCEP climatology, for 16 April 2012. (© Crown Copyright 2012, Met Office.)

anomaly component of this analysis for a single day. The information is critical for constraining air–sea exchange processes in NWP models and the synthesis of the underpinning satellite and *in situ* measurements provide the best of all worlds with regard to spatial coverage and calibration.

The changing composition of the atmosphere, be it of ozone in the stratosphere, nitrogen dioxide in plumes downwind of major cities, wind-borne dust or volcanic emissions, can also now be monitored using satellite-borne sensors. Natural and anthropogenic emissions of key chemical species may react with one another, with or without the presence of sunlight, or impact cloud microphysics through addition to the atmospheric condensation nuclei loading. Radiation and energy balances may thereby be disturbed as a result of changing atmospheric absorption and reflection characteristics. The resulting forcing can be sufficiently large for it to be important not just to assume 'climatological' atmospheric composition in NWP models. This incorporation of changing atmospheric composition into NWP has been dubbed *chemical weather* and it is enhancements in computing capacity which now make it possible to begin to couple meteorology and chemistry in a real-time way. Sensors continue to be developed and deployed, based on ultraviolet and visible spectroscopy, for example the Ozone Monitoring Instrument (OMI) on board the AURA satellite, monitoring total ozone, aerosol characteristics, nitrogen dioxide (NO_2) and sulfur dioxide (SO_2) amongst other species.

Ground-based remote sensing

We may naturally first think of satellites in terms of remote sensing systems but there is a whole range of technology based on the ground which is also now widely deployed to monitor the state of the atmosphere remotely. Perhaps the most familiar are radar systems which, operating in the microwave range, provide information about precipitation intensity by measuring the microwave backscatter signal of falling hydrometeors (Figure 3.18). Rainfall radar is often shown in television weather broadcasts to show the evolution of the precipitation field over recent hours and the NWP model output covering the forecast period to follow is often displayed in a style which mimics the radar presentation. Different types of hydrometeor (rain, drizzle, snowflakes, hail) have different scattering characteristics and incorrect assumptions about precipitation type can, therefore, lead to errors in the estimation of rainfall rate. Falling hydrometeors may also evaporate below cloud base before reaching the ground. These, coupled with a range of other effects such as the 'bright band' caused by any melting snowflakes (large scatterers) and the occultation caused by ground clutter, require that raw radar returns are interpreted with the assistance of NWP model data and generally corrected accordingly. There is no doubt, though, that radar is a very useful tool for supporting nowcasting over the next hours of the forecast period. Doppler radar is also increasingly

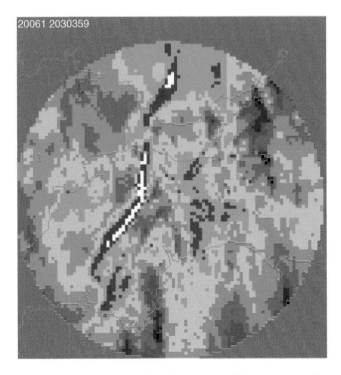

Figure 3.18   Rainfall intensity at 04:00 UTC on 3 December 2006 as measured by the Cobbacombe Cross radar covering a portion of South-West England and South Wales domain at 2 km resolution. (© Crown Copyright 2006, Met Office.)

being deployed in the developed world, providing additional information concerning the speed and direction of falling hydrometeors (Meischner, 2010).

Radar wind profilers and SODARs (Sonic Detection and Ranging) provide alternative means by which vertical profiles of (parts of) the troposphere can be derived, the former based on the radar principle and the latter dependent upon acoustic scattering. An increasing number of radar wind profilers in particular are now used to feed into NWP data assimilation and are sometimes supplemented by a co-located RASS (Radio Acoustic Sounding System) which provides information on the accompanying vertical temperature profile (Figure 3.19). An advantage of such systems over radiosondes is that they record continuously rather than being constrained to main synoptic hours at 00:00 and 12:00. Meanwhile, their vertical resolution in the lower troposphere out-performs what is currently possible from satellite.

Ground-based GPS receiver networks make it possible to estimate three-dimensional water vapour fields and total precipitable water through the delay in the transmission of the GPS satellite transmission caused by

Figure 3.19 Wind profiler and accompanying RASS deployed in Beijing. (Courtesy of Vaisala Oyj.)

atmospheric moisture (MacDonald *et al.*, 2002). Meanwhile, through ground-based lightning detection networks, such as the World Wide Lightning Location Network (WWLLN – http://webflash.ess.washington.edu/), lightning discharges are monitored using very low frequency (VLF) sensors. Real-time maps of lightning distribution provide a useful tool for monitoring severe weather and the accuracy of the forecast.

Summary

- Today's meteorological observing network has grown on a rather *ad hoc* basis to supply information for many different purposes.
- The current observing network is certainly not ideal for the purpose of providing data for the initialisation of NWP forecasts.
- Global coverage of meteorological data is dominated by satellite-based observations.
- A wide range of land-based observational networks, including radar, wind profilers, Doppler radar, meteorological buoys and radiosonde launching stations, provides a wealth of other data for meteorological purposes.
- Satellites do not measure meteorological quantities such as temperature or humidity directly, and sophisticated algorithms are necessary to convert satellite observed radiances into meteorological variables.

4

NWP Models – the Basic Principles

At the heart of any operational forecasting centre lies a numerical weather prediction (NWP) model. In fact, as we shall see in later chapters, it is more likely that large forecasting centres maintain more than one NWP model, and different models are used for different purposes. All NWP models are based on very similar principles, in an attempt to synthesise all the essential physical processes that shape the way the atmosphere develops in time. However, because of the independent nature of most state funded forecasting agencies, each major centre tends to maintain its own model or set of models, although most do tend to share their forecasts with agencies in other countries. There are also several examples of meteorological services in different countries using the same NWP model, but set up over a domain which is unique to that country. For instance, the HIRLAM regional model is used by 10 different European Meteorological services and the US WRF and MM5 models have been widely adopted by National Weather Services around the globe. As we shall see later, the variety of different models available adds a useful dimension to the task of forecast production. Because of the variety of different models that are currently used operationally it is sometimes difficult to describe a 'generic' NWP model, as there will always be differences between models. However, the basic principles are always similar.

In this chapter those basic principles are described, with discussion of how some of those principles are applied differently at different forecasting centres. This description falls into several parts. Firstly, the 'dynamics' of models is covered. This is essentially the representation of large scale motions of the atmosphere in response to large scale gradients of temperature and pressure, together with the Earth's rotation. Much of the dynamics of an NWP

Operational Weather Forecasting, First Edition. Peter Inness and Steve Dorling.
© 2013 John Wiley & Sons, Ltd. Published 2013 by John Wiley & Sons, Ltd.

model is concerned with the *conservation* of various quantities – momentum, mass, energy and water. Secondly, the 'physics' of models is examined. In NWP terms this basically means all the other processes that occur in the atmosphere on smaller scales but which are, nonetheless, essential to the evolution of the weather. Much of what is described as 'physics' in an NWP context pertains to *sources and sinks* of those same quantities of which the dynamical equations are describing the conservation. This division between *dynamics* and *physics* is a rather arbitrary one, since surely everything that happens in the atmosphere is really just applied physics. The division largely comes about due to the inherent scales of weather forecasting models as will be seen. Section 4.1 covers the dynamics and physics of an NWP model.

Once the basic physical principles have been decided upon, every NWP model needs a framework within which those principles can be applied to the model atmosphere in order to advance the forecast in time. Whilst all models include most of the same physical principles within their dynamics and physics, every model is unique in the way that those principles are applied in order to produce a weather forecast. This uniqueness largely comes about through the use of different frameworks and ways of solving the equations describing the underlying physical principles. Even within the same modelling centre, two different NWP models may include the same laws of physics, but these may be applied to the atmosphere and moved forward in time in rather different ways. This diversity largely comes about as a result of the main applications of the models – that is, the type of forecasts they are actually being used for. In Section 4.2 we will look at how the basic scientific principles are actually applied within NWP models to produce forecasts.

A further key principle of NWP is the need for a set of *initial conditions* from which to start the forecast. As discussed in Chapter 2, NWP is what mathematicians would call an *initial value problem*, so setting the initial conditions has a critical influence on the outcome of the forecast. Whereas all forecasting centres are reasonably in agreement on what are the important physical processes to include in an NWP model, there is less agreement between centres on how to go about setting the initial conditions of the forecast. All centres use a process called *data assimilation* to blend together the available observations of the state of the atmosphere with the laws of physics embodied within the NWP models. However, there are many different ways of doing this, different observational data sets are used at different centres and a wide range of different numerical techniques is available to blend them together to form the initial conditions. Section 4.3 examines some of the similarities and differences in data assimilation methods used in different forecast centres. This section will by no means be a comprehensive survey of the theory and techniques of data assimilation; several other excellent textbooks cover this topic in a great deal more depth.

4.1   The basic ingredients of an NWP model

To produce a forecast of the state of the atmosphere, the laws of physics which determine how that state will evolve in time need to be written down. The set of physical laws that govern the evolution of the atmosphere is actually rather small. Most of the fundamental laws that govern atmospheric motion concern *conservation* of key quantities – momentum, energy, mass and water. Within a completely closed system, principles of conservation of these quantities would be sufficient for production of a forecast. Of course, the atmosphere is not a closed system, as it interacts with the Earth's surface, both land and ocean. Energy is also constantly being exchanged across the top of the atmosphere, with solar radiation arriving and terrestrial and reflected solar radiation leaving. Over short periods – a day or so – and on reasonably large space scales, the sources and sinks of energy and momentum are reasonably small and, in fact, many early attempts at NWP ignored the sources and sinks or treated them in extremely simplified forms. The first ever computerised NWP forecast, run by the Princeton team on the ENIAC computer, actually produced a pretty reasonable forecast of the 24-hour evolution of the 500 hPa geopotential height field by using a scheme based on conservation. However, once one starts to produce forecasts for longer time ranges, or with more local detail, particularly near the Earth's surface, the sources and sinks become increasingly important. Of course, with increasing computing power, together with advances in our scientific understanding of the source and sink terms, all NWP models in operational use today include a comprehensive description of the important source and sink terms in their fundamental equations.

4.1.1   Dynamical equations

A list of fundamental scientific principles that govern the evolution of the atmosphere can be written down as a set of equations. These equations are often referred to as the *dynamical equations*, or the *model dynamics*. These names imply an emphasis on atmospheric motion, and that is indeed a key element of the dynamical equations. However, the conservation of energy is also an important element of the dynamical equation set, as is the conservation of mass. In fact, a set of five fundamental laws which describe between them the evolution of the atmosphere and make it possible to make forecasts of the future atmospheric state can be written down.

- **Newton's Second Law**. This essentially describes the conservation of momentum and explicitly allows the calculation of how the momentum of a body will change in response to an applied force or forces.

- **The Conservation of Mass** (often referred to as the principle of *continuity*). This law states that mass can be neither created nor destroyed. This principle breaks down at relativistic scales, as described by Einstein and others, but this is unimportant on atmospheric time and space scales.
- **The First Law of Thermodynamics**. This law describes the conservation of energy and, specifically, allows the calculation of how the temperature of a body may change as heat is added to it or work is done upon it.
- **The equation of state for a perfect gas**. This principle relates the temperature, pressure and density of a gas, allowing the calculation of any thermodynamic quantity given a prior knowledge of two other thermodynamic quantities.
- **Conservation of water vapour**. Water can be present in the atmosphere in all three phases – liquid, gas or solid – and can change easily between them. Of course, water is being constantly lost from the atmosphere in the form of precipitation and gained through evaporation from the Earth's surface, so, in terms of weather forecast production, this is perhaps the main principle in the list where the source and sink terms are of most crucial importance. In early NWP forecasts, water vapour was generally ignored altogether, partly because of the difficulty in quantifying the sources and sinks. Forecasts were made of the state of a dry atmosphere and then an experienced forecaster could use those forecasts to estimate where clouds and rain would form.

Newton's Second Law (the momentum equations)

Newton's laws of motion state that a body will remain at rest or in motion with constant momentum unless a force acts upon it. The second law states that when a force does act upon the body, the change in momentum will be directly proportional to the force. In atmospheric terms, if we change the word 'body' to 'parcel of air' and if, additionally, we make the assumption that all parcels of air have a mass of 1 kg, then we have a law that allows us to calculate the acceleration of air parcels if we can quantify the forces acting up on them. This is of profound importance in weather forecasting. If we can calculate all the relevant forces acting on parcels of air then we can calculate any changes to the speed of movement of air parcels – essentially allowing us to forecast the wind speed.

We can apply Newton's Second Law independently to motions in all three directions (west–east, south–north and vertically) in order to calculate accelerations in all three directions.

Mathematically, Newton's Second Law applied to air parcels of 1 kg can be written $\sum F = a$, where a is acceleration. The list of forces which can act on air parcels is rather short.

Pressure gradient force. This is the fundamental force in the atmosphere that sets air parcels in motion, or changes their momentum if they are already moving. Its magnitude is proportional to the local gradient of pressure and the force acts to accelerate parcels from higher pressure towards lower pressure. This includes vertical pressure gradients, which will always act to push air parcels vertically upwards.

Gravity. Acting vertically downwards towards the centre of the Earth.

Friction. This force dissipates momentum at the Earth's surface due to interaction of air parcels with surface roughness elements, such as plants, trees, buildings and so on. The effects of surface friction are translated upwards through the lowest part of the atmosphere by shear stresses – that is, air that has been slowed down by friction at the surface acts to slow the layer of air flowing immediately above it.

These are all 'real' forces in the sense that they are due to physical interactions between air parcels and their environment. We also need to consider the 'apparent' forces due to the fact that the Earth is rotating beneath the atmosphere as it moves, since Newton's laws of motion assume a frame of reference which is not itself accelerating.

Coriolis force. This is introduced to account for the apparent deflection of air motions over the Earth's surface because of the fact that the surface itself is moving (i.e. rotating). The force is proportional to the speed of motion of the air parcels and acts perpendicular to the direction of motion, resulting in an apparent deflection of air motions to the right in the northern hemisphere and to the left in the southern hemisphere.

Centrifugal force. This accounts for the apparent acceleration of an air parcel towards the axis of the Earth's rotation. For practical purposes this force is usually accounted for by combining it with the gravitational force in a variable known as *effective gravity*. Because true gravity acts towards the centre of the Earth but the centrifugal force acts away from the axis of rotation, effective gravity is not directed exactly towards the centre of the Earth. However, because the Earth is an oblate spheroid (i.e. flattened slightly at the poles and bulging slightly at the equator) effective gravity acts perpendicularly to a tangent to the local surface of the Earth.

We can now write down an equation for the motion of air parcels in each of the three directions:

$$\frac{du}{dt} - 2\Omega v \sin\phi + 2\Omega w \cos\phi + \frac{uw}{r} - \frac{uv\tan\phi}{r} = -\frac{1}{\rho}\frac{\partial p}{\partial x} + F_x$$

$$\frac{dv}{dt} + 2\Omega u \sin\phi + \frac{vw}{r} + \frac{u^2\tan\phi}{r} = -\frac{1}{\rho}\frac{\partial p}{\partial y} + F_y$$

$$\frac{dw}{dt} - \frac{u^2 + v^2}{r} - 2\Omega u \cos \phi = -\frac{1}{\rho}\frac{\partial p}{\partial z} - g + F_z$$

u, v and w are the three components of the wind in the west–east, south–north and vertical directions, respectively;

p is atmospheric pressure and ρ is atmospheric density;

Ω is the Earth's rotation rate, ϕ is latitude and r is distance from the centre of the Earth.

The terms on the right-hand side of each equation are the real forces, with the pressure gradient term being proportional to the rate of change of pressure along the component direction, and the friction terms (F) always acting against the motion in each component direction. Gravity appears only in the vertical momentum equation.

On the left-hand side of each equation, the d/dt term is the acceleration of the motion in that component direction. The terms in Ω (the Earth's rotation rate) are the Coriolis terms for the motion in that direction. The remaining terms (all proportional to $1/r$, where r is distance from the centre of the Earth) are there to account for the spherical geometry of the Earth. Consider, for instance, motion of an air parcel that is parallel to the Earth's surface at a particular point. If that motion continues in a straight line, at some distance away from the original point the air parcel will be further above the Earth's surface simply due to the curvature of the Earth beneath its trajectory. Hence, there will have been an apparent acceleration of that parcel in the vertical.

Conservation of Mass

Following a parcel of air along its trajectory, the mass of that parcel, M, cannot be changed, although its shape and volume may vary.

Mathematically:

$$\frac{dM}{dt} = 0$$

Expressing this in terms of the parcel's density and volume gives the continuity equation:

$$\frac{1}{\rho}\frac{d\rho}{dt} + \left[\frac{\partial u}{\partial x} + \frac{\partial v}{\partial y} + \frac{\partial w}{\partial z}\right] = 0$$

First Law of Thermodynamics

This law is a statement of the conservation of energy. Any heat added to a parcel of air is used either to increase the temperature, T, of the air parcel or to do work against the surrounding atmosphere as the parcel expands, or a combination of the two.

This is often expressed meteorologically as:

$$Q = c_p \frac{dT}{dt} + \alpha \frac{dp}{dt}$$

where Q is heat, c_p is the specific heat capacity of the air and α is the specific volume of the air ($= 1/\rho$).

Meteorologists often prefer to use *potential temperature*, θ, as a temperature variable rather than actual temperature, T. Defined as $\theta = T(p_0/p)^{R/c_p}$, where p_0 is a reference pressure level (1000 hPa), the potential temperature is the temperature a parcel of air would have if brought adiabatically to the reference pressure level. By definition then, potential temperature does not change as air parcels move adiabatically to higher or lower pressures, so the term in the thermodynamic equation which accounts for temperature changes due to expansion or compression of the air can be eliminated.

The *Lagrangian* rate of change of temperature of an air parcel can be expanded out into its component parts to give an *Eulerian* expression involving the advection of temperature by the winds. At a fixed point in space, the rate of change of any variable with time can be decomposed into changes due to air parcels with a different value of that variable being moved past the point by the winds (i.e. advection) and changes in the value of that variable within the parcels themselves due to sources or sinks. Expressing this mathematically:

$$\frac{\partial F}{\partial t} = \frac{dF}{dt} - u\frac{\partial F}{\partial x} - v\frac{\partial F}{\partial y} - w\frac{\partial F}{\partial z}$$

where $\partial F/\partial t$ is the local rate of change of any meteorological quantity (F) at a fixed point in space, dF/dt is the change in that quantity following individual air parcels (i.e. the Lagrangian rate of change) and the terms such as $u\partial F/\partial x$ express the advection of F by the local winds, since $\partial F/\partial x$ is the local gradient of that quantity in the west–east direction and u is the local wind component blowing across that gradient.

This means that we can write the thermodynamic equation in terms of potential temperature changes at a fixed point in space as:

$$\frac{\partial \theta}{\partial t} = -u\frac{\partial \theta}{\partial x} - v\frac{\partial \theta}{\partial y} - w\frac{\partial \theta}{\partial z} + \frac{\theta}{T}\frac{Q}{c_p}$$

Q represents the combined sources and sinks of heat.

The equation of state for a perfect gas

This is a *diagnostic* relationship between the fundamental thermodynamic variables – that is, there is no time derivative involved. Knowing any two thermodynamic quantities of a parcel of air (pressure, temperature, density, potential temperature etc.), the equation of state allows the calculation of any other thermodynamic variable. In meteorology the usual form of the equation of state is:

$$p = \rho RT$$

where R is the gas constant for dry air (= 287 J kg^{-1} K^{-1}).

Conservation of water

Water vapour in the atmosphere is usually expressed in terms of a mixing ratio q (i.e. mass of water vapour per unit mass of dry air). Following an air parcel the conservation of water can be written as:

$$\frac{dq}{dt} = E - C$$

where E refers to sources of water vapour (evaporation) and C refers to sinks (condensation). As with the First Law of Thermodynamics, the Lagrangian rate of change can be expanded into its various Eulerian components to give an expression in terms of changes in water vapour at a fixed point due to advection of water vapour by the wind field together with sources and sinks:

$$\frac{\partial q}{\partial t} = -u\frac{\partial q}{\partial x} - v\frac{\partial q}{\partial y} - w\frac{\partial q}{\partial z} + E - C$$

This complete set of equations clearly implies a set of fundamental *prognostic* variables which the NWP model will forecast. These variables are the three wind components u, v and w, a pressure-related variable (e.g. pressure on a height surface or *geopotential height* on a pressure surface), a mass-related variable (e.g. density), a temperature variable (e.g. T or θ) and a moisture-related variable (e.g. water vapour mass mixing ratio q). NWP models can produce forecasts of many other variables (e.g. cloud amount, rainfall rates) but these seven are the bare minimum needed to advance the equations forward in time.

This set of equations represents an *almost* closed system, with the sources and sinks of the various quantities being predicted (i.e. momentum, heat and moisture) meaning that some further calculations will need to be performed to obtain a complete solution. Most of the equations include time derivatives which allow the equations to be integrated forwards in time, which is, of course, the ultimate goal of numerical weather prediction. If at some initial time we have information about the wind distribution, the mass distribution, the temperature distribution and the water vapour distribution in the atmosphere, we can use this equation set to calculate how these fields will change with time:

We can use information about the pressure distribution to calculate the
pressure gradient terms in the momentum equations. We can then use
these equations to calculate how the winds will change in time.

We can then update the density (mass) distribution using the continuity
equation.

The thermodynamic equation can be used to calculate the changes to the
temperature distribution.

Since we have calculated new states for the density and temperature fields,
we can use the ideal gas equation to calculate the new pressure distribution.

The water vapour equation also allows us to update the state of the water
vapour field.

We then work through this process again, using the updated fields to calculate
further changes to each of the meteorological variables. For each step we
advance the equations forward over a finite length of time, known as the
model *time-step*.

Approximations to the equations

Over the years since the inception of NWP, a number of different approximations have been applied to the earlier equations. Some of these approximations
have been made simply to save computing time or simplify the solutions of
the equations. Other approximations have been made in order to suppress or
filter out undesirable, non-meteorological solutions of the equations. Modern
NWP systems tend to make fewer approximations to the equations than their
early counterparts, as increasing computing power together with advances
in the mathematical methods used to solve the equations have rendered
some of the approximations unnecessary. Here, some of the more common
approximations and their effects are looked at briefly.

'Shallow atmosphere' approximation In the equations of motion, the curvature terms include r, the distance from the centre of the Earth. Since the
Earth's radius is 6.37×10^6 m, and weather forecast models are dealing
with heights above the Earth's surface of up to a few tens of kilometres, the
atmosphere itself is very shallow compared to the Earth's radius. Hence, r
can be replaced by the constant a, the Earth's radius, in the curvature terms
with little loss of accuracy.

'Traditional' approximation The Coriolis terms in $cos\phi$ and the curvature
terms which *do not* include $tan\phi$ in the momentum equations are generally
small compared to the other terms and so are sometimes neglected. This,
combined with the shallow atmosphere approximation, constitutes what has

become known as the 'traditional approximation'. This approximation is becoming less widespread as increases in computing power mean that the small overhead saved by making this approximation is outweighed by the increased accuracy of the equations.

Hydrostatic approximation In the vertical momentum equation, there is a very close balance between the force of gravity acting to accelerate parcels of air down towards the surface and the vertical pressure gradient force acting to push parcels of air vertically upwards. This is no coincidence of course, since it is gravity that sets the vertical distribution of air density, and hence pressure, so it is quite natural that these two forces should come into equilibrium. The hydrostatic approximation assumes that these two forces balance so closely that any vertical acceleration will be negligible, so the $\partial w/\partial t$ term is zero. This is a reasonable approximation to make on scales of motion of 10 km or more, and even holds in many cases at smaller scales than this. Many NWP models have used the hydrostatic approximation over the years and, in some cases, this has been necessary in order to maintain the numerical stability of the models, since the hydrostatic approximation eliminates vertically propagating sound waves, which may amplify and become numerically unstable. In recent years, models have been developed, and are continuing to be developed, which attempt to explicitly resolve motions on smaller horizontal space scales (of order 1 km in some cases), so the hydrostatic approximation is not an appropriate one to make in these models.

Models which make the hydrostatic approximation but no further approximations are known as *primitive equation* models. These models still allow horizontally propagating sound waves as a solution.

Anelastic approximation This approximation involves neglecting the time derivative of density in the continuity equation. This effectively states that three-dimensional divergence of air must be zero, or that horizontal divergence must be balanced by vertical motion which acts to maintain the air density. This approximation filters out horizontally propagating sound waves as a solution to the equation set, since these waves require three-dimensional divergence in order to propagate.

Quasi-geostrophic approximation In the quasi-geostrophic approximation, winds are approximated by their geostrophic (i.e. exactly proportional to the pressure gradient) values in the acceleration terms in the momentum equation and the advection terms in the temperature equation. Vertical acceleration is neglected (i.e. the hydrostatic approximation is made). The quasi-geostrophic equations filter out sound waves, gravity waves and inertial oscillations. The original NWP forecast made on the ENIAC computer (Section 1.1) was based on the quasi-geostrophic system. However, even at the time, Jule Charney acknowledged that the system had limited application for weather forecasting, since it could not describe small scale motions of importance to weather forecasting, such as mesoscale circulations in frontal zones (Charney *et al.*, 1950). He argued strongly for the development of the *primitive equations* as the basis for NWP. After all, Richardson's original hand-calculated NWP forecast had been based on the primitive equations rather than the much simpler quasi-geostrophic set. That forecast became numerically unstable largely because high frequency gravity waves were permitted as solutions and these grew rapidly to swamp the forecast. Charney realised that if this problem could be overcome then any real advances in numerical weather forecasting techniques would come through the use of the more accurate primitive equations rather than the highly approximated quasi-geostrophic system.

4.1.2 Physical parametrizations

As stated above, the system of equations is *almost* closed. There are source and sink terms in several of the equations, which mean that the equations themselves do not represent a complete description of atmospheric processes. Most early NWP systems ignored these source and sink terms completely. In fact, because many of the early models were producing forecasts of the state of the mid-troposphere (at heights of about 5 km above the Earth's surface) out to lead times of about 24 hours, this was not too bad an approximation. Since many of the sources and sinks of momentum, heat and water vapour are near the surface, these terms are often rather small for the mid-troposphere over short periods. However, in all modern NWP systems we want to produce forecasts for the whole atmosphere, including the part nearest the surface, and we want forecasts for periods of longer, and sometimes considerably longer, than 24 hours. Hence, these sources and sinks need to be quantified and equations for their prediction need to be written down. The method of doing this is known as *parametrization*. Effectively, the physical processes are not being explicitly represented, but their impact on the various model variables is determined through a set of simplified calculations which make use of a range of *parameters*. Some of these parameters can be determined from physical experiments and *in situ* measurements, while others are highly empirical and are chosen largely through trial and error within NWP models to see which values give the best forecasts. Parametrization involves a great deal of pragmatism. There may be some complex physical processes which are understood very well, but to represent them explicitly within an NWP model would take too much computing time and so the processes need to be simplified in order to include their effects. There may be other processes which are not fully understood, but if they have a significant effect on the state of the atmosphere they still need to be included within our models. In these cases, the way the processes are represented may not be particularly physically realistic but this is acceptable if there is a positive impact on the quality of the forecasts.

In each of the basic equations which contain source and sink terms (momentum, heat and humidity), there may be multiple physical processes acting as the sources and sinks. It may also be the case that a single physical process may act as a source or sink in more than one of the equations. Atmospheric convection is an obvious example of a process that acts as a local sink or source of all three of these fundamental quantities. Hence, it makes more sense to discuss the source and sink terms by atmospheric process.

Radiative transfer

The fundamental source of heat to the Earth/atmosphere system is electromagnetic radiation from the Sun. Though the energy from the Sun spans a wide range of the electromagnetic spectrum, the majority of the energy arriving at the top of the atmosphere is in the visible part of the spectrum. However, small but important amounts of energy at much shorter wavelengths than this also arrive from the Sun and need to be accounted for within forecast models.

The energy arriving from the Sun at the top of the atmosphere is largely a function of the surface temperature of the Sun and the Earth–Sun distance. These things are well known and so the spectrum of solar energy entering the atmosphere can be easily defined. What happens to the radiation as it passes through the atmosphere is much more complex, as the energy can be transmitted, reflected, absorbed or scattered by the various different gases and aerosols present in the atmosphere, by clouds and by the Earth's surface itself. Different wavelengths of radiation will also behave in different ways. For instance, the short, ultraviolet wavelengths of solar radiation are strongly absorbed by ozone (O_3) present within the stratosphere whereas longer wavelengths (in the blue part of the visible spectrum for instance) are hardly absorbed by the atmosphere at all but are more likely to be scattered by various atmospheric constituents, accounting for the blue colour of the sky. Much of the *absorption* of solar radiation occurs at the surface of the Earth and the rate of absorption is strongly dependent on the nature of the surface. Fresh snow and ice absorbs very little solar radiation, reflecting about 90% of it back up into the atmosphere whereas dark surfaces, such as black soils or the ocean surface, absorb much of the radiation that falls upon them. The absorption of sunlight by the oceans is further complicated by the solar elevation. At low elevations, such as just after sunrise and just before sunset, the low incident angle of the solar radiation on the ocean surface means that most of the

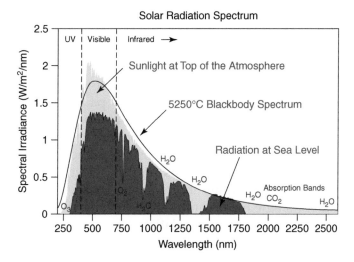

Solar Radiation Spectrum

Figure 4.1 Spectrum of solar radiation at the top of the atmosphere (yellow shading) and at the Earth's surface (red shading). The idealised spectrum for a black body at 5250°C is shown by the grey curve. The chemical symbols show the main absorbing gases

radiation is reflected. Figure 4.1 shows the spectrum of solar radiation at the top of the atmosphere (yellow shading) and the same spectrum at the Earth's surface (red shading). Clearly the interaction between solar radiation and the atmosphere is rather complex and also very wavelength dependent. This complexity needs to be represented with NWP models, but if it is done so in a rigorously explicit, wavelength-by-wavelength way, forecasts would take too long to compute, so a considerable degree of simplification is required.

The Earth's surface and the atmosphere itself also emit radiation, although at much longer wavelengths, and hence with much less energy than the Sun. Nevertheless, this radiation represents a very important part of the radiation budget of the Earth–atmosphere system. The greenhouse effect, for instance, whereby infrared radiation from the Earth's surface is absorbed by various atmospheric constituent gases (largely carbon dioxide and water vapour, but also methane, nitrous oxide and ozone) accounts for a large warming of the Earth's climate, so is also a crucial element of an NWP forecast model. Emission of energy in the infrared part of the spectrum by the Earth's surface is the main mechanism for reducing the surface temperature overnight, and hence this mechanism has a strong determining influence on the forecast of overnight minimum temperature.

Because there is very little overlap between the radiative spectrum of the Sun and that of the Earth and its atmosphere (because of the huge difference in their surface temperatures), the two components of the radiation budget can be treated completely separately within numerical models. This is usually done by splitting the radiative parametrization into *shortwave* (i.e. solar) and *long wave* (i.e. terrestrial) parts. Within these two spectra, the radiation is usually broken down into a set of different wavelength *bands*, and the fate of the radiation in each band is then treated separately by the parametrization scheme. Even using fairly broad spectral bands instead of doing wavelength-by-wavelength calculations results in radiative parametrizations that are computationally very expensive. To reduce this overhead, radiation schemes are often not called at every single time step, with the calculations from a radiative transfer time step being held constant (or adjusted only for time of day) over the subsequent time steps until the next full calculation. An alternative strategy is to compute the radiative transfer on a reduced grid rather than at every single model grid point, with the values at intermediate grid points being calculated by interpolation.

There are two main groups of atmospheric constituents that affect the transmission, absorption, scattering and emission of radiation. The first is the various constituent gases in the atmosphere. The mixing ratios of these gases are largely fixed for a particular temperature and pressure, and very precise laboratory experiments have been performed which determine the particular properties of each constituent gas at any given wavelength of radiation. The spectral properties of each constituent gas are held in look-up tables so that their radiative effects can be efficiently computed. The one atmospheric gas that varies considerably in both time and space is water vapour, and the mixing ratio of water vapour at each grid point and time step is calculated by the model itself.

The second group of atmospheric constituents that affects the transmission of radiative energy through the atmosphere is clouds. The radiative effects of clouds are much more complex than those of a clear-sky atmosphere, and it is in the treatment of clouds and their interaction with radiation that NWP model radiation schemes differ most widely. Clouds are fractal in nature, with variations on all space and time scales that can affect how they interact with any incident radiation. The size and shape of the individual cloud particles (liquid or solid) will affect the way that they absorb or scatter radiation and even extremely thin layers of cloud can have significant radiative effects. The impacts of all of this variability need to be included in any radiative parametrization; this presents a major challenge to model development.

Shortwave radiation   The spectrum of incoming solar radiation at the top of the atmosphere can be calculated very precisely as a function of time of year, time of day and latitude. Typically, the solar spectrum is broken down in parametrization schemes into between two and ten finite bands. In some simpler schemes the ultraviolet and visible parts of the spectrum are treated together and the near-infrared band is treated separately. In more sophisticated schemes the ultraviolet is dealt with separately from the visible part of the spectrum and the visible spectrum itself may be split into several bands. The fate of the solar radiation as it passes through the atmosphere is then computed from the *radiative flux divergence equation*. This equation considers the incoming radiation at the top of each layer of atmosphere and then computes how much of that radiation will be scattered, absorbed and transmitted in the layer, and hence how much radiation will emerge at the bottom of the layer, as input into the layer below. These factors will depend on the mixing ratios of various atmospheric constituent gases in the layer, the temperature of the layer and, crucially, the presence and nature of any clouds in the layer. The treatment of the interaction of radiation with clouds in an NWP model is an extremely complex process and many different schemes exist which attempt to quantify this. At the bottom of the atmosphere the interaction of solar radiation with the surface is also very important, particularly for determining the evolution of the temperature of the surface and the lowest layers of the atmosphere. The nature of the surface at each grid point determines the partitioning between reflection and absorption of the incident radiation. The *albedo* of the surface at each grid point is specified in advance, usually based on satellite or *in situ* measurements. The incident radiation that is reflected back up to the atmosphere then needs to be treated in the same way as the incoming radiation.

Of course, it is through absorption that radiation is able to modify the temperature of the atmosphere or the Earth's surface, and hence act as a heat source. The radiative heating rate for a particular layer in the atmosphere will be proportional to the amount of radiation absorbed within the layer, which itself is calculated as the residual from the radiative flux divergence calculations discussed above:

$$\left(\frac{\partial T}{\partial t}\right)_{rad} = -\frac{g}{c_p}\frac{\partial F}{\partial p}$$

where F is the radiative flux.

Long-wave radiation The treatment of long-wave radiation is similar to that of shortwave radiation. Once again the long-wave spectrum is split into a number of finite width spectral bands. These are chosen to represent parts of the spectrum with similar characteristics, such as the so-called 'atmospheric window' regions of the spectrum between about 8 and 9.7 μm and 11–12.5 μm, wavelengths to which the atmosphere is largely transparent, and the water vapour band between 5 and 8 μm, a region of the spectrum where water vapour absorbs strongly. Because of the greater variations of absorption with wavelength in the infrared (Figure 4.2) the long-wave spectrum is usually dealt with in more bands than the shortwave.

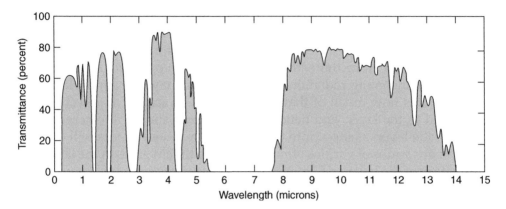

Figure 4.2   Atmospheric transmittance of electromagnetic radiation. Note the band between 5 and 8 μm where water vapour absorbs very strongly

The major difference between the treatment of short and long-wave radiation is that the atmosphere and Earth's surface *emit* significantly in the long wave part of the spectrum. The energy emitted will be proportional to the fourth power of the temperature of the surface or layer of atmosphere, as determined from Stefan's Law.

Once again, in order to act as a heat source in the forecast equations, the long-wave radiation needs to be absorbed by the atmosphere. This absorption depends on the presence of particular absorbing species and on the wavelength of radiation. The heating rate will be proportional to the radiative flux divergence.

Surface and boundary layer processes

The main sink of atmospheric *momentum* is surface friction. The amount of friction varies with the nature of the Earth's surface, so any representation of this effect will be an inevitable approximation since clearly the effect of every hill, building and vegetation element cannot be represented. The effect of a sink of momentum at the surface will be a downward flux of momentum through the boundary layer due to shear stresses exerted by each layer of the atmosphere on the layer above. Turbulent and convective eddies in the boundary layer also transport momentum vertically. The magnitude of the momentum flux and the depth of the atmosphere through which it extends need to be calculated by the model at each location and time step.

The first requirement for these calculations is a surface flux condition at the lower boundary. This is usually represented as being proportional to the product of a drag coefficient and the square of the wind velocity in the lowest model level. The drag coefficient is a function of the nature of the surface at every location within the model domain and will also be a function of the atmospheric stability in the surface layer.

Once the flux at the lower boundary has been determined the flux of momentum through the rest of the boundary layer needs to be calculated. This is often done though a method known as *K*-theory. This theory uses an analogy with molecular diffusion where a flux of a quantity such as heat or momentum through a fluid layer will be proportional to the gradient of that quantity in the layer. The constant of proportionality is the diffusion coefficient for that quantity in the particular fluid. *K*-theory treats macroscopic parcels of air in boundary layer eddies as if they were individual molecules, and thus calculates fluxes which are proportional to the vertical gradient in the boundary layer. The constant of proportionality, K_M, is known as the *eddy diffusivity* for momentum and will clearly have larger values than the analogous molecular diffusion coefficient. The value of K_M will depend largely on the stability of the atmosphere within the boundary layer. In very stable conditions vertical overturning in eddies will be suppressed, so K_M will take on a low value. In unstable conditions boundary layer eddies will be of greater vertical extent, so K_M will become large.

The *closure* of a parametrization scheme is the way in which the grid-box scale variables are used to diagnose the magnitude of the subgrid scale fluxes. The use of *K*-theory-based schemes to represent turbulent vertical mixing is usually referred to as using a *first order closure*. This is because only information about the *mean* vertical gradient (i.e. the first statistical moment) of a particular quantity is used to diagnose the turbulent fluxes of that quantity. Higher order closures are increasingly being used within NWP models as increases in computing power allow more sophisticated schemes to be included. *Second order closure* involves using the subgrid scale *variances* of quantities such as temperature, moisture and momentum and the vertical wind component to calculate vertical fluxes more explicitly (variance is the second statistical moment). This approach involves the solving of many more equations and the need for the model to carry more information from time step to time step, so is still not a realistic approach in most operational NWP applications. An intermediate approach is to use what is known as a *one-and-a-half order* closure. This retains the mean gradient approach of *K*-theory for most calculations of turbulent transport but also involves solving an equation to predict the *turbulent kinetic energy* (TKE) of the flow. TKE is a measure of the high frequency variance of the wind components and is a useful quantity for determining how effective turbulence is at mixing the various quantities being predicted by the model. This approach is commonly used in high resolution NWP models in which turbulent mixing by boundary layer eddies is an extremely important factor for accurate prediction of near-surface properties of the atmosphere, such as the evolution of surface temperature through the diurnal cycle or mixing of pollutants through and out of the boundary layer.

In lower resolution and global NWP models, the first order approach is more often used. This approach is often extended by a consideration of *non-local mixing*. This is based on the idea that, in some stability conditions, boundary layer eddies have a large vertical extent and can mix quantities such as heat, moisture and momentum very rapidly through the whole depth of the boundary layer. The vertical mixing is thus calculated by considering not only the local gradients at each model layer but also by examining the gradients throughout the boundary layer. Particularly in unstable conditions the mixing is enhanced by the non-local effect.

The transfer of *sensible heat* between the Earth's surface and the lowest layer of the atmosphere (or vice versa) is usually calculated as being simply proportional to the temperature difference between the surface and the lowest model level, with a constant of proportionality known as a *transfer coefficient* – a function of the atmospheric stability. The calculation of vertical turbulent transport of heat through the boundary layer is dealt with in a similar way to that of momentum, with the transport by turbulent eddies being proportional to the vertical gradient of temperature, with an eddy diffusivity parameter, K_H, as the constant of proportionality. In some parametrization schemes the value assigned to K_H will be the same as that for K_M, the eddy diffusivity for momentum. In more sophisticated schemes these values are calculated separately. In either case, the eddy diffusivity is calculated as a function of the local atmospheric stability and becomes larger as stability decreases. Sensible heating rates in the boundary layer are calculated as a residual of the sensible heat flux divergence in a model layer (i.e. the difference in sensible heat flux between the bottom and the top of the layer).

One part of the surface heat budget that is not perhaps immediately obvious, but which can have a significant impact on the forecast of surface temperature, is the flux of heat into or out of the ground beneath the surface. NWP models all include a ground heat flux model to account for this effect. Usually the upper 10 metres of so of the solid Earth are divided up into a number of layers which become increasingly thick with depth. Heat fluxes are then calculated based on the vertical gradient of temperature in the soil, together with some transfer coefficients which are usually dependent on soil type.

By far the main source of water vapour, and hence *latent heating*, to the atmosphere is the evaporation of water from the Earth's surface. This process is relatively easy to deal with over an open water surface such as an ocean. Evaporation rates are proportional to the humidity gradient between the surface (assumed to by saturated at the temperature of the surface) and the lowest model layer, multiplied by the wind speed in the lowest model layer. The transfer coefficient will be a function of stability in the surface layer. The situation becomes more complicated over land surfaces. If the soil is not completely dry there will still be some evaporation from a land surface and, in addition, water held on vegetation (e.g. on the leaves of trees) is also potentially available to the atmosphere through evaporation. Furthermore, all vegetation draws water up through its roots and loses it to the atmosphere via transpiration from the *stomatal pores* on the leaves. Transpiration rates are complex functions of soil moisture, vegetation type, temperature, atmospheric humidity and time of day. All of these factors can affect the evolution of atmospheric temperature and humidity, so need to be accounted for within NWP models. Errors in soil moisture can account for quite large errors in surface temperature forecasts in NWP models due to incorrect partitioning of the incoming solar radiation between warming up the ground surface and evaporating moisture.

Further complication arises because most model grid boxes consist of a mix of different surface types, such as trees (broad or needle leaf), grassland, shrubs, bare soil, urban areas and open water. All of these different surfaces have different capacities for holding water at the surface, and will have different transfer coefficients for evaporation and transpiration. Some NWP models attempt to account for this variability by specifying a *surface tiling scheme*, whereby the surface of each grid box contains a fractional value of each of the various surface types. Surface fluxes are calculated for each tile separately and then combined as an area- weighted average for the grid box as a whole. A *resistance*-based method is often used to calculate fluxes of water from vegetated surfaces, whereby the flux to the atmosphere is determined by the gradients of humidity between the surface and the atmosphere, scaled by a number of different resistances to moisture transfer. These resistances represent the various physical parts of the systems, such as the drawing-up of moisture through the plant roots or the transpiration of water from the leaves into the atmosphere.

Modelling the effects of evaporation and transpiration can be done in a whole range of different ways, involving different levels of complexity. The level of complexity used will depend on a number of different factors, such as available computing power, but also upon the type of forecasts which are actually being produced. An NWP model being used to predict surface temperature evolution on a very finely detailed grid may include a great deal of complexity in these calculations, whereas a global domain NWP model with large grid boxes may use much simpler calculations with no real detriment to the accuracy of the forecast.

Once water has been evaporated from the surface its transfer upwards through the boundary layer can be handled in much the same way as that of sensible heat. Vertical fluxes in subgrid scale eddies will be proportional to vertical gradients of humidity, scaled by an eddy diffusivity for water vapour. This exchange coefficient will be a function of atmospheric stability and is usually taken to be the same as the eddy diffusivity for sensible heat. At the top of the boundary layer, humidity can be exchanged with the free atmosphere through entrainment/detrainment. This exchange will itself be stability dependent.

Convection

Above the boundary layer, the main vertical transport of both sensible and latent heat is achieved in convective processes. There are many different ways of parameterizing convection in NWP models, so the convective parametrization tends to vary widely between different models. The grid scale of an NWP model will also have a big influence on how convective processes are dealt with. In a model working on a grid scale of tens or even hundreds of kilometres, convective processes will all be much smaller than the grid itself, so have to be dealt with entirely within a parametrization scheme. Once the grid scale gets below 10 kilometres, some convective process start to be resolved (i.e. represented on the scale of the model grid), so there is an additional complication over which processes to parameterize and which processes will be explicitly represented by the equations discussed in Section 4.1.2.

The convection scheme of a numerical model also has to do more than simply calculate heating rates due to transport and absorption of sensible and latent heat. Convective cloud amounts need to be calculated; these will then feed into the radiation scheme. Convection also acts as a major sink of atmospheric water vapour through precipitation and, of course, in terms of weather forecast production, most customers are more interested in precipitation than they are in atmospheric heating rates.

In early numerical models of the atmosphere, all of which ignored the presence of water within the atmosphere, the role of the convective parametrization was simply to prevent the atmospheric columns becoming absolutely unstable in the vertical so that large scale vertical motions occurred on the scale of the grid. Simple 'convective adjustment' schemes acted to redistribute heat in the vertical to prevent absolute instability. The first convection scheme in a model which included moisture was the *moist convective adjustment* scheme of Manabe *et al.* (1965). This scheme simply redistributed heat in the vertical so that an unstable profile was moved towards a moist adiabat. Over tropical oceans, the vertical temperature profile is often observed to be close to a moist adiabat so this parametrization, although extremely simple, works pretty well over large parts of the globe which exhibit regular convective activity. However, it works far less well over land regions where convective available potential energy (CAPE) is often observed to build up in a profile over many hours before convection occurs, resulting in vertical profiles which are far from moist adiabatic. As NWP models became more sophisticated and started to include water substance and radiative transfer schemes, so convective parametrizations had to become more sophisticated too, dealing with vertical transports of water and calculations of cloud amounts.

In many numerical models today, convective processes are dealt with through a 'mass flux' approach, such as that described by Tiedtke (1989) and used in the ECMWF forecast model. In each atmospheric column that exhibits a degree of instability to convection, a convective mass flux is calculated which is proportional to some measure of instability within the column. Typically, this measure may be the convective available potential energy in the column. This mass flux then forms the basis of a convective plume or *updraught* that rises from one model level to the next, carrying sensible heat and water vapour and liquid/frozen water as it rises. At each level, the mass flux may be altered by the entrainment of air from outside the convective plume or detrainment of cloudy air into the environment. The temperature of the updraught will be adjusted to allow for adiabatic expansion; condensation of water vapour will also adjust the temperature of the rising plume. A similar technique is applied to negatively buoyant convective downdrafts driven by cooling due to evaporation of precipitation falling through the unsaturated part of the grid column. Both updraught and downdraught act to transport heat in the vertical and stabilise the atmospheric column.

At each level of the atmospheric column, the temperature of the air in the rising plume is adjusted for all of the entrainment, detrainment, condensation and evaporation processes. The buoyancy of the rising plume is then tested by comparison with the temperature of the large scale environment, and a set of criteria are applied to see whether the plume will continue to rise or whether it will terminate. Some convective parametrizations, such as those by Arakawa and Schubert (1974), deal not with a single plume but with an ensemble of plumes within each grid column. These plumes are assumed to have a distribution of buoyancy, so, at each vertical level, a proportion of the plumes may become neutrally buoyant, whereas the rest can still be positively buoyant.

Many convective parametrizations perform some simple preliminary tests, such as an undilute parcel ascent, in order to determine whether convection in each column at each time step will be deep penetrating convection extending into the free troposphere, shallow convection limited to the boundary layer or mid-level convection which initiates above the boundary layer. Different parameters, such as the rate of entrainment of environmental air into the convective plume and buoyancy criteria, are then applied given the particular convective circumstances.

One problem with complex convection schemes like the mass-flux approach is that it is almost impossible to measure quantities such as convective mass fluxes within real clouds, so it is very hard to say what the mass flux profile *should* really look like. Add to this the further difficulty of measuring entrainment and detrainment into and out of a real convective plume and it becomes clear that convection is a process that has to be parameterized in a fairly pragmatic way, using values for various parameters that give the best results in tests of the forecast model. One approach that is regularly used to get around this issue is that extremely high resolution simulations of convection are run in models that can explicitly represent some of the small scale processes which are impossible to measure in real clouds. These simulations can then be used as verifying 'truth' for convective parametrization schemes.

Within a sophisticated convective parametrization, convection itself can affect the temperature of the large scale environment through:

- Detrainment of air from the convective cloud into the large scale environment due to mixing at the cloud boundary. Air from the convective plume will have a different temperature to the environmental surroundings, so detrainment will change the large scale temperature. This effect will be proportional to the mass of air detrained, which itself will be proportional to the total mass flux in the convective plume. Any liquid water detrained from the cloud is usually assumed to evaporate instantaneously, thus cooling the large scale environment.
- 'Terminal detrainment'. As the convective plume becomes neutrally or negatively buoyant and so stops rising it effectively 'dumps' the air from within the plume into the large scale environment. The same process will happen if a convective downdraft loses its negative buoyancy or reaches the ground. The detrained air mixes with the environment, thus altering the temperature. In most NWP models detrained liquid water (or ice) is assumed to evaporate, which will cool the environment. There is a growing trend, however, that some detrained water or ice goes immediately into forming *layer* cloud if the grid box humidity is high enough to support it.
- Compensating subsidence. As air rises within the convective plume, air outside the clouds must be sinking to compensate for the upward transfer of mass within the cloud. This air will warm adiabatically as it descends.
- Evaporation of falling convective precipitation in subsaturated air outside the clouds. This will cool the large scale environment. The same is true of melting of frozen precipitation falling through the subcloud layer. For the sake of simplicity, all of the cooling due to melting is usually assumed to occur in the layer which contains the zero degree C isotherm (usually referred to as the *melting level*).

Convection can also affect atmospheric heating rates indirectly through the presence of convective clouds which interact with shortwave and long-wave radiation within the radiative parametrization scheme.

Convective momentum transport Parcels of air rising within convective clouds carry heat and moisture vertically. They also have a value of *horizontal momentum* associated with them. As the parcels rise they will carry this momentum with them, usually to layers of the atmosphere with higher values of horizontal momentum. Hence, cumulus convection can act as a brake on the wind speed at the levels through which it extends; this will be particularly true if the convection is deep, extending from the boundary layer high into the free troposphere. However, momentum transport by cumulus clouds is very much more complex than heat and moisture transport. Whereas heat and moisture are approximately conserved within convective updraughts, momentum is generally *not* conserved. The degree of conservation of momentum depends very strongly on the vertical wind shear environment in which the clouds have formed. Many NWP models attempt to account for the effects of momentum transport by convection, and there are many different ways of doing so. These range from simple 'cumulus friction' parametrizations to more complex schemes which take into account the vertical shear in the cloud environment and even the pressure gradients that occur across the boundaries of the clouds.

Clouds and large scale precipitation

A convective parametrization deals with buoyancy-driven vertical motions occurring on scales smaller than the model grid. Vertical motion on scales at or above the model grid can also lead to the formation of clouds and precipitation, such as the large scale slantwise ascent of air within a frontal zone. Whilst the vertical motions themselves may be resolved by the model, the processes that determine the thickness and horizontal extent of the cloud and the growth of precipitation particles happen at scales much smaller than the grid, so still need to be dealt with through parametrization. In fact, one of the important controls on cloud and precipitation formation is what happens at the scale of individual cloud droplets or ice crystals – the so-called 'microphysics' scale. The formation and subsequent growth of cloud droplets is a complex result of the interaction of local humidity and the nature of the microscopic *cloud condensation nuclei* (CCN) onto which water vapour is able to condense. This gets particularly complex in the ice phase, where the shape and size of CCN have a very strong control on the formation and growth of ice crystals.

Very few operational models attempt to deal with this level of complexity in any rigorous manner as this would be computationally expensive. It is also the case that the physics of cloud droplet formation and growth are still only partially understood and extremely hard to measure, so a large degree of simplification is applied in most operational NWP models.

Many modern NWP models use what is called a *prognostic cloud scheme*. This means that cloud liquid water and cloud ice become fundamental model variables which are carried from time step to time step. They have their own predictive equations and are advected by the large scale flow. This is an advance on *diagnostic cloud schemes*, which simply diagnose cloud amount at each time step as a function of the humidity in each grid box. Prognostic cloud schemes usually show a smoother evolution of the cloud field with time, with more continuity in the cloud structure within weather systems than diagnostic schemes, which can have a rather noisy evolution of cloud from time step to time step.

In both cases, the formation of new clouds is related to changes in the saturation specific humidity, often labelled q_{sat}. This is a function of temperature and pressure, so can be modified by adiabatic processes such as vertical motion and diabatic processes such as radiative cooling. The condensation rate in existing clouds is usually taken as proportional to the time rate of change of q_{sat}. Clouds grow if q_{sat} is decreasing with time and dissipate if q_{sat} is increasing.

New clouds are assumed to form when the relative humidity in the grid box exceeds some critical threshold value, RH_{crit}. The value of RH_{crit} is usually specified in advance for each model level; it is usually around 80% but often higher for clouds in the lowest model levels. The fraction of grid box covered by cloud is itself usually a function of RH_{crit}, with higher values of RH_{crit} leading to larger cloud fractions. Although 80% relative humidity may seem too low to lead to cloud formation, in a grid box that may typically cover 500 km² the subgrid scale fluctuations in relative humidity would almost certainly mean that some fraction of the area would be saturated if the area mean was 80%. Whether cloud forms as liquid or ice, or a mixture of both, is usually taken to be a simple function of temperature, with all cloud above an upper threshold temperature (usually 0°C) being water, all cloud below a lower threshold being ice, and all cloud forming at temperatures between the two thresholds being a mixture of water and ice.

If microphysics is dealt with within the cloud scheme, it is usually in the context of conversion of cloud water/ice and precipitation between different states. Figure 4.3, taken from Wilson and Ballard (1999), shows schematically the various conversion processes in a typical cloud and precipitation parametrization. Equations can be written down for all of the processes shown, with temperature being a key variable in determining conversion rates between the different states of water and ice in the grid box.

A crucial part of the microphysical calculations is to calculate how much precipitation will fall at each time step. For liquid clouds, some parametrizations attempt to represent, in a very simplified and often statistical way, the growth of cloud droplets into drops which are able to fall out of the cloud under the influence of gravity. Within water clouds the main growth process

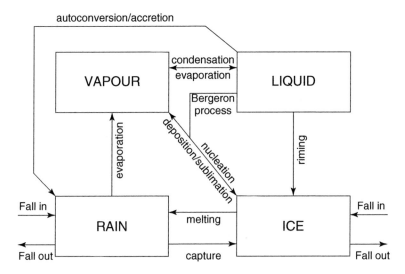

Figure 4.3 The microphysical processes represented in a typical cloud and precipitation parametrization. (Adapted from Wilson and Ballard (1999), by permission of John Wiley & Sons, Ltd.)

is collision with other cloud droplets; the *drop size distribution* within the cloud controls how efficiently this process occurs. Other cloud schemes use a very simple *autoconversion* method, whereby a certain fraction of the cloud liquid is assumed to fall out as rain. This fraction is often partly controlled by a *critical cloud liquid water content*, above which precipitation is assumed to become more efficient. This critical value is sometimes loosely determined by observational studies. Autoconversion is often also used to calculate the frozen precipitation (i.e. snowfall) rate in ice clouds, although usually at a higher rate than for liquid precipitation.

In the mixed or ice phases of clouds, precipitation growth can be very complex. For instance, in mixed phase clouds some parametrizations allow frozen precipitation to grow more quickly in order to crudely model the Bergeron–Findeisen process, whereby cloud ice grows at the expense of cloud liquid water.

Once precipitation falls from a cloud layer it interacts with all the layers beneath through which it falls. Rain and snow falling through cloud-free, sub-saturated air may partially or totally evaporate; precipitation falling through cloudy air will grow by condensation and by collision and coalescence with cloud droplets/ice. The final precipitation rate at the ground will be due to the integrated effect of all of these processes in the column above.

As well as predicting precipitation rates and cloud amounts as model output variables, a crucial role of a cloud scheme is to predict cloud amounts as input into the radiation parametrization. How the radiation scheme actually 'sees'

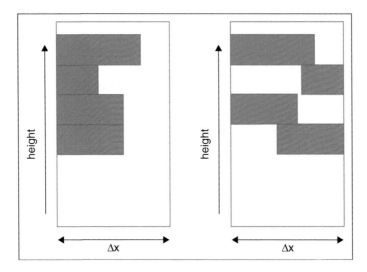

Figure 4.4 Schematic representation of two grid columns from an NWP model with horizontal dimension Δx. Both columns contain four identical cloud layers. In the left-hand column these layers are maximally overlapped and there is still some clear sky in the column. In the right-hand column the layers are minimally overlapped and there is no clear sky in the column

clouds has a critical impact on the energy budget of the atmosphere and so can materially affect the evolution of the forecast.

A number of assumptions have to be made to simplify the representation of clouds so that they can be treated within the radiation scheme in a tractable way. The first simplification made in models is in the shape of clouds. Anyone who has spent any time observing clouds will have been struck by their continuously changing and fractal nature. In NWP models, however, clouds are represented as plane-parallel slabs, occupying a fraction of each grid box. The next assumption that needs to be made is in how clouds in different model layers within a grid column are overlapped in the vertical. Clearly this will have a big impact on how much solar radiation gets through to the ground. If different cloud layers in a column are overlapped as minimally as possible, the grid column could appear to be totally cloudy when viewed from above, whereas if the different cloud layers are maximally overlapped then there will still be some clear sky in the column viewed from above or below (Figure 4.4). In most cloud parametrizations it is assumed that clouds in adjacent model levels will be maximally overlapped as they are probably representing different levels of the same cloud. If there are one or more clear layers between two layers containing cloud, then the amount that they overlap is usually assigned randomly, as these layers may have no physical association with each other. Such a set-up is referred to as 'maximum random overlap'.

Vertically propagating gravity waves

In stable atmospheric conditions, mountains and hills generate internal gravity waves that propagate vertically with increasing amplitude. At some upper level these waves 'break', destroying their horizontal momentum in turbulent motions. This effect will be very familiar to anybody who has flown in an aircraft over a major mountain range and encountered turbulence. Gravity wave breaking acts as a sink of momentum at the level at which it occurs, so the effect needs to be accounted for in NWP models. All major operational NWP models include a *gravity wave drag* parametrization which is associated with orography. The speed of the flow, atmospheric stability and vertical wind shear are all input parameters to the parametrization.

Deep convection can also generate vertically propagating gravity waves which break in the stratosphere, so convective gravity wave drag parametrizations are now starting to be developed and tested in operational forecasting models.

4.2   Building the physical principles into a model

As they stand, the physical principles that determine how the atmosphere evolves in time can be written down as equations. These equations include time derivatives that allow them to be used to produce weather forecasts. The methods used for applying these physical principles are as important as the principles themselves in the process of weather forecast production. Forecasters would like to apply the physical principles as accurately as possible to the atmosphere, but it must also be possible to produce numerical forecasts rapidly, so that they can be of some use to the customers who will be using them. These two requirements are often conflicting, so building an NWP model which meets both criteria poses a big challenge to the scientists and programmers working on model development teams. In this section we will look at some of the choices that have to be made when applying the basic scientific principles of atmospheric development in an operational NWP context. Many of these choices relate to the mathematical methods which are used to integrate the model equations forwards in time. These methods can be extremely complex and we will largely avoid discussion of this complexity. Other texts exist which cover this in much more detail. Instead, we will focus on an overview of the problem – how do we use an extremely complex system of equations in a practical way in order to produce weather forecasts which are both sufficiently accurate and available in the required timeframe?

4.2.1   Finite difference methods

The first important point about the governing physical principles of atmospheric development is that the equations themselves have no analytical solution which can be written down and applied at any point in the future. The only way to use the equations to produce weather forecasts is to advance them forwards in finite steps of time, replacing the continuous derivatives in space and time with some sort of finite approximations. The best way to understand this is to consider an example – in this case the momentum equation for motion in the east–west direction:

$$\frac{du}{dt} - 2\Omega v \sin\phi + 2\Omega w \cos\phi + \frac{uw}{r} - \frac{uv \tan\phi}{r} = -\frac{1}{\rho}\frac{\partial p}{\partial x} + F_x$$

This equation contains one time derivative (the du/dt term) and one spatial derivative (the $\partial p/\partial x$ term).

Looking at the time derivative first. du/dt is the rate of change of zonal momentum (u) with time, *at an infinitesimal point in time*, and if we can find values for all the other terms in the equation, we would be able to find out precisely how the value of u was changing *at that instant in time*. However, if we want to discover how much u will change at any finite time in the future we would have to do an infinite number of calculations, which is clearly impractical for operational weather forecasting. What is done in NWP models is to specify a finite *time step*, call it Δt, and state that over that time interval the value of u will change by a finite value, which we might call Δu. We then replace the du/dt term with its 'finite difference' approximation, $\Delta u/\Delta t$. As long as we choose a value of Δt which is not too long then this is a reasonable thing to do. However, how do we know what constitutes 'not too long' and how can we quantify the inaccuracy that we introduce by replacing a derivative by a finite difference approximation? We will address this a little later.

Now consider the spatial derivative, $\partial p/\partial x$. This is the *gradient* of pressure in the east–west direction at a fixed point in space. Can we quantify this at every point in space? If the pressure was a simple mathematical function of, say, latitude and/or longitude then yes, we could write down a value for this derivative everywhere. For instance, if pressure only varied as the sine of distance in the east–west direction (i.e. $p = sinx$) then we would know that $dp/dx = cosx$. However, the atmosphere is never that simple. The pressure can never be specified as a simple, differentiable mathematical function, so here too we need to resort to using a finite difference approximation to specify dp/dx. If we know the values of p at two fixed points in the atmosphere, separated by a finite distance, Δx, then we can write an approximation to dp/dx as $\Delta p/\Delta x$, where Δp is the difference in pressure between the two points. Once again, as long as Δx is 'not too big' this is a reasonable approximation. So as long as we know the pressure at lots of fixed points in space, covering our area of interest, we can write down values for the pressure gradient anywhere within that area.

This brings us to one of the fundamental principles of building a numerical model – that of using a three-dimensional grid to represent the state of the atmosphere. The atmosphere can be divided up into lots of contiguous three-dimensional boxes and in each box the atmosphere is assumed to have a single, unique value of each of the fundamental prognostic variables described in Section 4.1. These boxes, usually referred to as *grid boxes*, are usually cuboid in shape, but this does not have to be the case, and NWP model developers are currently working on models with hexagonal grid structures and even grids which can change size and shape as the model runs, so as to continually re-focus computational resources where they are most needed. Figure 4.5 shows a schematic picture of an NWP global model grid, with the atmosphere divided up into many boxes, with layers of these boxes in the vertical.

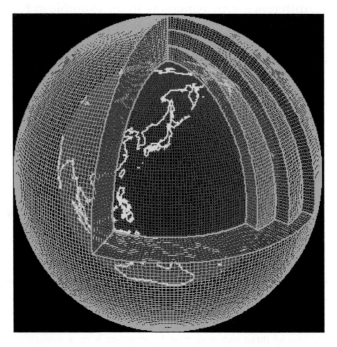

Figure 4.5   A schematic representation of grid boxes in a global numerical model of the atmosphere. Each box holds a single value of each of the prognostic variables of the model

This figure gives the impression that the vertical division of the atmosphere is done in concentric spherical shells, or layers like an onion. In fact, most NWP models use a vertical co-ordinate that follows the terrain of the Earth's surface rather than being based purely on height above sea level or distance from the centre of the Earth. The reason for doing this is rather obvious. Real air parcels flowing around the atmosphere near the surface will flow over obstacles such as large mountains rather than flowing through them, so it makes sense for the model levels to follow the terrain rather than pass through it. A model vertical co-ordinate based purely on height above sea level would have some grid boxes which were actually below the ground. In many NWP models the vertical co-ordinate is designed to be completely terrain-following near the surface and completely horizontal high in the atmosphere (above, say, 10 km) with a blending between the two systems in between. This means that parcels moving along a model vertical level near the surface will flow over the hills, air parcels in the middle part of the troposphere will still feel an influence of the orography below them, but air parcels high up at the tropopause level will move along fairly level surfaces which are not influenced by any orography.

Returning to the problem of specifying the zonal pressure gradient (the dp/dx term) we can see that we could use the value of pressure at the two grid boxes either side of the box for which we are trying to specify the gradient to give us the finite difference approximation. In the example in Figure 4.6, if we are trying to find the value of dp/dx in box **B** then we could use the pressure values in boxes **A** and **C** to do this.

Writing this down mathematically:

$$\left[\frac{\partial p}{\partial x}\right]_B \sim \frac{p_C - p_A}{2\Delta x}$$

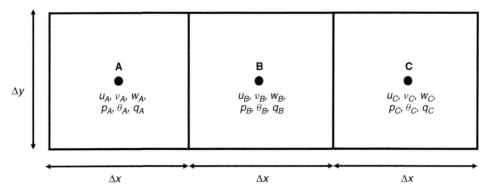

Figure 4.6   A plan view of three two-dimensional grid boxes from a numerical model. Each box holds a single value of pressure (p), potential temperature (θ), the three vector wind components (u,v,w) and specific humidity (q) in the centre of the box

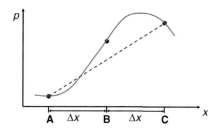

Figure 4.7 A possible distribution of pressure across grid boxes A, B and C is shown in the solid line. The finite difference approximation to this distribution is shown in the dashed line

This is the simplest possible approximation to the pressure gradient that we could make, and actually it is not a very good approximation at all. Figure 4.7 shows one possible true distribution of pressure across boxes A, B and C in the solid line, together with the finite difference approximation of the gradient, calculated using the earlier equation, in the dashed line.

Clearly, in this case, the finite difference approximation underestimates the pressure gradient at **B** considerably. One possible solution would be to make the grid spacing, Δx, much smaller, but if we halve Δx then the model has to perform twice as many calculations to produce the same (albeit more accurate) weather forecast. There would be no point in halving the grid spacing just in the x direction – we would want to do the same in the y (north–south) direction and this would make the model four times slower (and, therefore, more expensive) to run. In fact, as we shall see later, halving the grid length in a model generally also requires halving the time step, introducing even more calculations and making the model eight times slower to run.

One way of getting around at least some of the inaccuracy of the finite difference method without introducing an unmanageable number of extra calculations is to hold different prognostic variables at different points of the grid box. In the example of the pressure gradient calculation above, we are calculating dp/dx at **B** in order to calculate the time rate of change of u at **B**. If, therefore, we held the values of p at the *edges* of the grid boxes and the values of u (and also the other wind components) at the *centre* of the grid boxes, as shown in Figure 4.8, then we could approximate the pressure gradient by

$$\left[\frac{\partial p}{\partial x}\right]_B \sim \frac{p_C - p_A}{\Delta x}$$

This is more accurate because we have reduced the horizontal distance across which we are calculating the gradient from $2\Delta x$ to Δx, but without doubling the number of calculations the model has to perform.

Both of these two calculations are called *centred difference approximations* to the gradient, as they are based on looking at the values either side of the grid

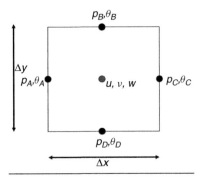

Figure 4.8 An NWP grid box in which the wind variables are held at the centre of the box and the pressure and potential temperature are held at the edge of the box

point in order to calculate the gradient at the grid point. In meteorological models, when we know that the wind is blowing from a certain direction, it might make more sense to calculate the gradient on the *upwind* side of the grid point, since it is from this direction that the air is being advected. This method then becomes numerically unstable, so more mathematically sophisticated methods need to be used to suppress the instability. This moves into a realm of numerical methods which is beyond the discussion in this book.

4.2.2   *Numerical stability*

Let us return to the question of what constitutes 'too long' for an NWP model time step, or 'too big' for a model grid spacing. It should be fairly clear from the above discussion that the shorter the model grid length, the better will be the approximations for calculating gradients using finite difference methods. The same is also true for model time steps – a shorter time step implies a more accurate calculation of rates of change. There are no upper limits for the length of time step or the size of the grid boxes within a numerical model, but it is essential that these two quantities are *consistent* with each other.

Consider what would happen if the grid spacing in an NWP model were reduced by say four or eight times without changing the model time step. In the original configuration air parcels may have been able to move from one grid box to an adjacent one in the duration of a single time step. However, with a much reduced grid size, air parcels would be able to move across several grid boxes in one time step. In this case, the finite difference method of using the *local* gradients of, say, pressure or temperature to calculate rates of change becomes inappropriate, since it is the gradient across the entire path length of an air parcel's motion that becomes the appropriate gradient to determine the time rates of change.

A consequence of this is the fact that, in the reduced grid length situation, the calculations of rates of change of the prognostic variables will become *numerically unstable*, or *unbounded*, allowing these variables to grow rapidly and unrealistically in time. Over the course of a few time steps quantities such as the wind speed can grow so large that the model 'blows up'. It can be shown that, in order for a finite difference scheme to remain numerically stable, the time step must be shorter than the grid length divided by the fastest moving signal in the atmosphere. Another way of saying this is that air parcels must not be able to move across more than one grid box during one time step. This is known as the *Courant–Friedricks–Lewy (CFL) condition*, which can be expressed mathematically as:

$$\Delta t \leq \frac{\Delta x}{c} \quad or \quad c\frac{\Delta t}{\Delta x} \leq 1$$

where c is the fastest possible atmospheric motion in the model. So, if we assume that the fastest motions in the atmosphere are of the order of $100\,\mathrm{ms}^{-1}$, then a model with a grid spacing of 50 km would need to have a time step of less than 500 seconds, whereas a model with a grid spacing of 5 km would need to have a time step of less than 50 seconds.

The CFL criterion for numerical stability can cause problems in numerical models near the poles. If latitude and longitude are used to define the model grid (an obvious choice) then the longitudinal grid length will get smaller towards the poles due to the convergence of the meridians. Several approaches are possible to deal with this problem.

The CFL condition is not universally applicable in all numerical schemes that are used to integrate the model equations forward in time. Numerical schemes do exist that allow longer time steps to be used in a numerical model than would be suggested by application of the CFL condition. This is advantageous because it makes the model quicker and cheaper to run, although not necessarily any more accurate. Such 'implicit' or 'semi-implicit' schemes are designed to slow down the fastest moving waves within the solutions to the dynamical equations so that they do meet the stability criteria. Although this is physically unrealistic, it is acceptable within NWP since the fastest moving gravity waves usually have little meteorological significance. The mathematical details of these numerical schemes are beyond the scope of this text (Kalnay (2003) provides a more detailed discussion).

Another possibility that allows for a longer time step than suggested by the CFL criterion is the use of 'semi-Lagrangian' methods of integration. In a fully *Lagrangian* model, instead of calculating changes in meteorological variables on a fixed grid, individual air parcels are tracked as they move around the atmosphere. Whilst this is an excellent approach for a model being used for predicting, for instance, pollution or volcanic plume dispersal, it does have problems for a general forecast model, not least of which may be that all the individual parcels may bunch up in particular areas, meaning that some parts of the atmosphere have a very dense coverage of information whereas other areas may contain very few parcels. Semi-Lagrangian methods still work on a model grid, but the advection schemes calculate where the air parcels arriving at each grid point *would have come from* at each time step. By calculating these so called 'departure points', the scheme can then take into account the meteorological variables at the departure points when updating the values at each grid point, getting away from the purely local gradient information in a pure finite difference calculation at the grid point. This means that even if air parcels are travelling across several grid boxes during a time step, the meteorological information at their departure point is accounted for, so the numerical scheme remains stable. A method such as this, which allows longer time steps, means that the computing costs of running high resolution models can be kept down.

4.2.3   Grid box size and 'resolution'

Meteorologists often refer to the 'resolution' of an NWP, meaning something about the size of weather features that the model is capable of representing explicitly. A model with very small grid boxes will have a higher resolution (i.e. will be able to explicitly represent smaller features) than a model with large grid boxes. In fact, it is quite common to hear meteorologists use the terms 'resolution' and 'grid box size' entirely interchangeably. This is, in fact, quite misleading. We will look at a simple example to see why.

Many weather features have a wave-like form; this is perhaps clearest when we look at maps of atmospheric pressure or geopotential height. As well as closed circulations corresponding to depressions and anticyclones, we also see a lot of ridges and troughs with sinusoidal shapes. Figure 4.9a shows an idealised variation of surface pressure with horizontal distance in the x-direction. The variation is sinusoidal with a ridge of high pressure shown between the two vertical lines.

How many grid points do we need in the x-direction in order to capture the shape of this pressure distribution? In other words, how many grid points will allow us to *resolve* the ridge? Clearly we need more than one point in the region between the two vertical lines in Figure 4.9a in order to capture the shape of the feature, but how many more?

Figure 4.9b shows a situation in which there are three evenly spaced grid points across the ridge feature.

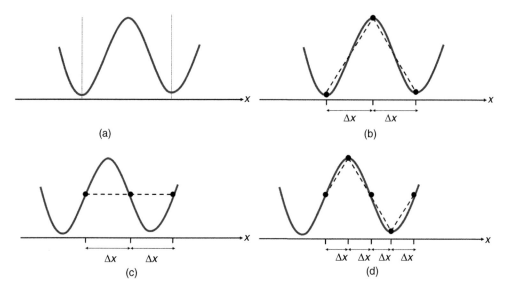

Figure 4.9   (a) A sinusoidal distribution of a meteorological variable along the x-direction. (b) The finite difference approximation to the sinusoidal distribution given three grid points evenly spaced in the x-direction is shown by the dashed line. (c) The finite difference approximation to the sinusoidal distribution if the three grid points are shifted by one quarter of a wavelength along the x-axis is shown by the new dashed line. (d) With five grid points within one wavelength of the sinusoidal distribution, the wave will be represented on the grid regardless of the position of the grid points relative to the wave

Having one grid point at each of the 'turning points' of the ridge means that the feature is just resolved by the model grid, although the pressure distribution seen on the model grid is saw-tooth in pattern rather than sinusoidal (as shown by the dashed line). Having the three grid points placed as they are means that the model knows that the ridge feature exists. However, what happens if the ridge moves along the x-axis by quarter of a wavelength but the three grid points remain fixed where they are? This situation is shown in Figure 4.9c.

Now we have a situation whereby the same three grid points each have the same value of atmospheric pressure, so that the model now knows nothing about the presence of the ridge and the pressure distribution 'seen' by the model is a completely flat one, as shown by the dashed line. This means that the ability of the model to resolve the presence of the ridge is dependent on the position of the ridge. Clearly, in order for the model to resolve the ridge regardless of where it is in space, more grid points (i.e. a smaller grid spacing) are required. Figure 4.9d shows a situation in which Δx has been reduced so that there are now five grid points spanning the wavelength of the ridge.

In this case the ridge will be resolved whatever its position along the x-axis, although in most cases the turning points of the pressure distribution will not correspond exactly with the position of the grid points, so the amplitude of the ridge will generally be underestimated.

This example shows that grid spacing and resolution are not the same thing, although the second does depend on the first. It is probably true to say that the *resolution* of an NWP model is four or five times the grid spacing, so that a model with a grid spacing of 10 km can just about resolve features which are 40 or 50 km in size. This is something of a generalisation, as it will also depend on the shape of the weather feature. However, it is important to recognise that model resolution is not exactly the same as model grid size.

4.2.4   Spectral methods

The finite difference approach is not the only possible way of describing the physics of the atmosphere in a numerical model. A radically different approach was pioneered during the 1970s by, amongst others, Brian Hoskins and Adrian Simmons at the University of Reading in the United Kingdom (Hoskins and Simmons, 1975). This approach is known as the *spectral method* and is used in a number of operational NWP models. In Section 4.2.1 it was stated that is not possible to write down the spatial distribution of atmospheric pressure (or any other meteorological field for that matter) as a simple mathematical function, gradients of which could then be calculated analytically. However, in very simple terms, this is what the spectral method attempts to do. Visual inspection of a map of meteorological variables such as pressure or geopotential height often shows that these fields have sinusoidal distributions in space, with troughs and ridges of different wavelengths being the major features to which the eye is drawn. The same is also true for fields such as temperature above the boundary layer, where effects such as the presence of orography and land–sea contrasts are small. The spectral method builds on this observation by decomposing meteorological fields into a series of *spectral harmonics* in two dimensions in much the same way as a Fourier series decomposes a one-dimensional set of data such as a time series into a series of sines and cosines. By choosing a set of spectral harmonics with a suitable set of different wavelengths, any atmospheric field in a numerical model can be represented as the sum of these harmonics.

Choosing the wavelength of the smallest spectral harmonic in the series is in some ways equivalent to choosing the grid length of a grid point model, as this choice determines what features will be resolved by the numerical scheme of the model. Spectral models are generally labelled by the wave number of the shortest wave in the set of spectral harmonics. Hence, a T799 model has its series of spectral harmonics truncated at a wavenumber of 799.

This means that 799 wavelengths of the smallest harmonic in the series would be able to fit around the equator. Dividing the length of the equator by 799 tells us that a model with this truncation would be able to resolve wave-like features with a wavelength of about 50 km.

The advantage of this method is that spatial gradients of atmospheric quantities can be computed analytically rather than having to use a numerical finite difference approximation, so are theoretically more accurate. A disadvantage is that the numerical methods needed to integrate a spectral model are extremely complex, although this is of course not an insurmountable problem given that a sufficiently skilful team of mathematicians is available to work on the problem. Another issue is that the spectral method is based on the continuous nature of the Earth's atmosphere (i.e. it has no edges in the horizontal), so this method cannot be applied within a model which is only representing a limited area of the globe. In terms of applying physical parametrizations in a spectral model, the model variables need to be transformed back onto a spatial grid in order to calculate the increments due to the different schemes, before being transformed back into wave space for the next set of dynamical calculations.

The global NWP models used by NCEP, ECMWF, Meteo-France, the Japanese Meteorological Agency (JMA) and the US Navy are all spectral models, whereas the UK Met Office has always used grid point models for its global NWP forecasts. Simply choosing a spectral model does not guarantee more accurate weather forecasts. The quality of the initial conditions for the forecast and the nature of the physical parametrizations will have considerable impacts on the forecast accuracy and are probably the first order factors that determine the accuracy of weather forecasts.

4.3   Setting the initial conditions for the forecast

As Chapter 2 made clear, an accurate specification of the atmospheric state at the start of the forecast period is as essential an ingredient of accurate weather forecasts as the correct specification of the laws of physics governing atmospheric motions, or accurate numerical methods used to apply those equations within the framework of an NWP model. Accurate specification of the initial conditions of a weather forecast is a hugely complex undertaking for several reasons. Firstly, the variety of different types of observational data means that a large amount of pre-processing of data has to occur before it can be used in determining the initial state for the forecast. Secondly, the mathematical challenge of blending a large set of observations into a complex numerical model without destabilising the state of the atmosphere is considerable. Thirdly, the huge volumes of observational data, particularly from satellites, make the exercise a considerable challenge in purely data handling terms.

In this section the methods used for generating initial conditions for an NWP forecast, often described as meteorological *data assimilation*, are examined. Which observations are used to create an initial state for numerical weather forecasts are looked at, how exactly these observations are used is considered and future developments in this area are discussed.

4.3.1   Combining models and observations – the philosophy of data assimilation

Chapter 3 showed that the global meteorological observing network has grown in a rather *ad hoc* way, in some cases over a period of more than 100 years. It is clear that the surface observing network (manned and automatic weather stations, radiosonde launch sites, ocean buoys and weather ships) was not designed to provide an optimum network for providing data for NWP forecasts. In densely populated parts of the developed world we have a very dense coverage of observing stations whereas some parts of the globe, such as the oceans and parts of the developing world, have very little *in situ* observational coverage at all (see, for instance, Figure 3.3). In many cases the regions with sparse observational networks are precisely the regions in which the weather systems of most interest to forecasters actually develop. Satellite-based observations clearly provide a more uniform observational coverage but these observations also have their limits. For instance, satellite soundings of temperature and humidity have rather coarse vertical resolution compared to a radiosonde.

An interesting question, therefore, is to ask what observational network we would design if we were doing so specifically to provide data for accurately setting the initial conditions for NWP forecasts. We could then look at how far this is from the observing network we actually have and consider how we can minimise the impacts of any differences between the ideal and the actual network. In order to do this, the first thing we need to consider is exactly what information an NWP needs in order to start producing a forecast. This is a pretty straightforward question to answer. The model needs a value of all the fundamental 'prognostic' variables (u, v, w, T or θ, p and/or geopotential height and q) at every single point on the model grid, including in the vertical. Hence, our ideal observing network could be composed of a radiosonde station in every single grid column of the model grid. It is pretty clear that this is totally impractical, particularly once we start getting down to models with a grid spacing of only a few kilometres. For models with a global domain this requirement would mean that we needed to launch many thousands of radiosondes over the oceans. If we consider the first of these problems, it is not obvious that we do need an *in situ* observation in every grid column of an NWP model if the grid boxes are only a few kilometres square. In many situations, the weather conditions in adjacent grid columns will be very similar, especially above the boundary layer, so such a dense network of observations will have a degree of redundancy built in for much of the time. A counter argument to this is that the occasions when conditions do change significantly across short horizontal distances, such as in the presence of a weather front, are precisely those conditions when we would like to have a very fine scale observing network. We clearly do, however, need meteorological data over the oceans, as many important weather systems such as tropical cyclones, intense mid-latitude depressions and polar lows all form over the oceans. Even if we did have observations of all the right variables at all the model grid points, there would still be potential problems with the *consistency* of those observations. Due to errors with the observing equipment, slight time differences between the recording of the observations, and different types of instrument being used to make the observations in different locations, it is quite probable that the observed data may not be consistent with each other. For instance, the wind values at a point should be consistent with the pressure field in the surrounding area. However, an instrument error in recording the wind or the pressure (or both), could mean that the observed pressure field was not consistent with the observed wind field. A consequence of this would be that, when the model started to run the forecast, the first thing it would try to do would be to bring the two fields back into balance by propagating high frequency gravity waves through the atmosphere. These waves would propagate outwards from the source of the inconsistency like ripples on a pond after a stone has been thrown in. This would result in some very strange weather patterns in the vicinity of the waves and may even result in the forecast becoming numerically unstable and 'blowing up'.

Although the existing observational network is clearly very far from our idealised case, there does exist a source of *exactly* the right meteorological variables on *precisely* the correct gridded network that we need for initialising an NWP forecast. That source is the forecast from the *previous run* of the same NWP model. For instance, if we are starting a forecast at 06:00 UTC on a particular day then the T+6 forecast from the 00:00 UTC run of the same model will include all the right variables on all the right grid points for our needs. Better still, all of these variables will be physically consistent with each other, since the laws of physics which form the model equations will have predicted a physically realistic evolution of the state of the atmosphere. Hence, the wind distribution will be exactly consistent with the pressure distribution, which in turn will be physically consistent with the temperature distribution. The obvious problem with using a short-range forecast from a previous model run to set our initial conditions is that, even just 6 or 12 hours into the forecast period, the forecast will *always* have diverged from reality to some extent. Sometimes this divergence will be large, more often it will be rather smaller, but it will always be non-zero and, as we have already seen, it is essential to set the initial conditions for a numerical weather forecast as precisely as possible.

The above discussion makes it fairly clear that the best way to set the initial conditions for a weather forecast is to use a blend of the short-range forecast from the previous run of the model and the observations that have been made in the intervening time in order to shift the model-predicted evolution back towards reality. This blending of model and observations is called *data assimilation* and it is not unique to weather forecasting. Many branches of science that include a modelling and an observational component (e.g. predicting oil reserves from geological surveys) use data assimilation methods to produce the best possible estimate of the state of a physical system at a particular time given incomplete data. Data assimilation is an eminently sensible way of producing the initial state for a weather forecast, since it combines a numerical model that includes all the relevant laws of physics which govern the evolution of the atmosphere together with observations which know nothing about those physical laws but which do have information about the state of the atmosphere at specific times and locations. The process of data assimilation is not simple and can involve some rather complex mathematics, due in no small part to the vast amount of data that needs to be handled and processed. However, it is a crucial part of weather forecast production. The importance placed upon data assimilation by operational forecast centres can be seen in the amount of high performance computing time that those centres devote to the data assimilation process in each forecast cycle. Whilst this varies from centre to centre it is universally true that more computing time is spent on data assimilation than is spent on actually running forecasts into the future.

Several different approaches to the data assimilation process have been used operationally over the years (e.g. 'successive corrections', 'nudging', 'optimal interpolation'). A number of different approaches are still being used operationally but, in general, most operational centres are now using either *three-dimensional or four-dimensional variational data assimilation* methods. The discussion here is focused on these methods. The mathematical complexity of these methods is not discussed in detail here.

4.3.2   *Variational data assimilation*

All data assimilation is designed to combine actual observations with a numerical model to ascertain the best possible estimate of the state of a physical system (in this case the atmosphere) at a fixed point in time or through a particular period known as the assimilation window. *Variational data assimilation* attempts to do this in a statistically rigorous way by minimising a *cost function* which includes several elements.

The first of these elements is the difference between the model 'background' trajectory through the assimilation window when uncorrected by observations and the new trajectory once the observations are incorporated. Basically this means that the data assimilation process does not try to push the model trajectory too far away from its original path because the model is applying the laws of physics to the atmosphere, so we do not want to end up with an atmospheric state which is inconsistent with the laws of physics. However, we know that there are errors associated with the model due to simplifications in the physical parametrizations and numerical errors introduced by the ways that the equations are solved, so these errors need to be taken into account during the assimilation process.

The second element that variational data assimilation attempts to minimise is the distance between the observations and the new model trajectory through the assimilation window. This is because the observations are the best measure of 'truth' that we have, so we want to make the best possible use of them. However, we do know that there will be errors associated with the observations due to instrument inaccuracies, the interpolation of observational data onto the model grid and translation of one type of observational data into a different meteorological variable (e.g. turning satellite-measured radiances into temperature and humidity profiles). Variational data assimilation takes the statistics of these errors into account.

The final thing that the variational assimilation approach attempts to minimise is the presence of high frequency, large amplitude, fast moving gravity waves. Introducing observations into a model field which is fully physically consistent is a little like throwing a handful of stones into a calm pond. The stones will introduce perturbations to the surface of the pond which spread rapidly outwards from the point of impact until eventually the surface levels off again. In an NWP model these ripples take the form of gravity waves as the model state adjusts rapidly to any spurious inconsistency between the mass and momentum fields. These waves can produce some rather unrealistic weather and can even cause model blow-up, so they are undesirable for accurate weather forecast production.

4.3.3 Observation processing

As already clearly shown in Chapter 3, the observations available to an NWP model come in many different types and formats. Prior to assimilation into the model, these data need to be processed into a consistent format that the model can read. This type of processing tends to be very specific to each forecast centre, as some centres use different formats for inputting observational data into their forecast models. However, the basic steps in each case are similar:

- Translation of the observation from its original coded format into a model-readable format. All meteorological observations arrive at a forecast centre in some form of numerical code, but these codes differ hugely between observation types. Hence, all the observations are processed into large files and written in a consistent form that the model can read and handle.
- Translation of observations into model variables. For example, wind observations from most deployed instruments will be in the format of direction (degrees from north) and speed, but NWP models use wind data in vector component (u,v) format. Another example is that screen measurements of temperature and humidity tend to be in the form of dry-bulb and dew point temperatures, whereas models tend to work with dry-bulb temperature and specific humidity.
- Translation of data from non-SI units. For historical reasons meteorological observing systems use a whole raft of different units for measuring different quantities. Wind speeds tend to be reported in knots (nautical miles per hour), visibility in metres but cloud heights in feet, temperatures in degrees Celsius and sometimes even in Fahrenheit. All NWP systems use SI units for physical consistency so all data in non-standard units need to be converted prior to assimilation.
- Checks that the accompanying metadata are in the correct format. For example, the vertical co-ordinate attached to a report of winds and temperature from an aircraft may be altitude above mean sea level, but the model may use pressure as the vertical co-ordinate. Observations with missing metadata (e.g. no location information) will be rejected.
- A crucial part of the observation processing is the assignment of *observational error statistics* to every observation to be used in the assimilation. Prior to any type of observation being used in NWP, a study will be made of the error characteristics of that instrument. This information is then used during the assimilation process in order to determine how much weight the observation should be given and how much it can be adjusted to fit better with other observations and with the model-predicted state. The error statistics may have a seasonal cycle and, for instruments reporting on multiple vertical levels, they will also have a vertical dependence.
- Data thinning. Several types of observation provide very fine spatial resolution data with many observed values per model grid box (e.g. satellite winds derived from cloud motion vectors; Figure 4.10). In such cases the excessive number of observations needs to be thinned to reduce the load on computing time.

- Blacklisting checks. Forecasting centres constantly monitor the impact that different types of observations have on the forecast skill. This is done by running parallel versions of the forecast in which different observational data sets are taken out of the assimilation process – so-called *data denial* tests. If a particular type of observation is found to have a systematic negative impact on the forecast skill it can be *blacklisted* so that it is automatically excluded from the assimilation. Further investigations can then be made off-line to discover the reasons for the degradation of the forecast. This may be due to badly calibrated instruments, instrument drift with time or various other factors. In one specific case, wind profiler data over North America were found to be degrading the forecast scores of the UK Met Office global model during the Autumn. The data were blacklisted and subsequent observations showed that, during this season, large flocks of migrating geese flying over the profilers were contaminating the wind observations!

Much of this processing is routine and rather obvious but, of course, without having the observations in the correct format they are unusable in an automated system, so this part of the operation of the forecasting system is as essential as the coding of the model equations themselves.

Once the observations have been processed into the correct format a rigorous set of checks must be applied to the observations to ensure that

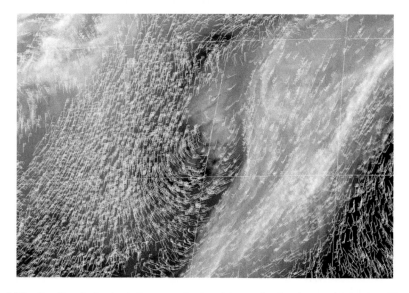

Figure 4.10 Satellite-derived wind vectors in the vicinity of a developing depression. The lines of longitude and latitude shown are drawn every 10 degrees. The colour of each vector indicates its vertical level. Yellow vectors are between 700 and 1000 hPa, white vectors are between 400 and 699 hPa and beige vectors are between 100 and 399 hPa. (© Crown Copyright 2004, Met Office.)

incorrect data are not being introduced into the process. There is a fine line to be trodden here, as observations which are unusual but correct should ideally be accepted as they may have an important influence on subsequent development of active weather systems. For instance, a ship at sea may be reporting a surface pressure value which is very much lower than its previous report and also lower than any other observations in the vicinity. This may be because an observer made an error reading the barometer or transmitting the observation, or it may be that the pressure is genuinely falling rapidly at that location – the precursor to a rapidly developing intense depression. Automated checking algorithms need to be able to distinguish between an erroneous observation and a genuine but unusual one. This is not always successfully achieved. On Boxing Day 1999 a major storm swept through Europe causing large amounts of damage to property, infrastructure and the environment. It was not particularly well forecast, even 24 hours in advance, by the ECMWF forecast model, partly because the automated observation checking routines had rejected genuine observations of low surface pressure from ships in the Atlantic as being outside the limits set by the checking algorithm.

Most operational schemes tend to err on the side of caution when checking the observations being allowed into the assimilation. Forecasting centres are prepared to accept occasional situations when good data are rejected as being too far from the model background in order to avoid frequent situations when data which are very different to the background trajectory are allowed into the assimilation and introduce instabilities into the model.

As well as checking against other observations, the main check on incoming data is to determine how far the observations are from the model background value (i.e. the value calculated by the model when run through the assimilation window with no observational corrections). Once again, this can lead to the rejection of true observations if the model has failed to predict the development of a rapidly deepening depression for instance. Therefore, the rejection limits need to be very carefully chosen and there will always be circumstances when correct observations are rejected.

A wide range of automated checks can be applied to observations before they are accepted for use in the assimilation. Some of these are independent of other observations, whereas others are dependent on observations in the surrounding area or in the same place but at different times. Most checks rely on knowing something about the variance or standard deviation of the variable under consideration. Clearly, rejection limits need to be larger on variables with large standard deviations. Some of the checks applied at operational centres are:

- Climatological checks. Does the observation fall outside its normal observed limits?

- Background departure checks. Is the observation too far from the model predicted background value?
- Buddy checks. Does the observation disagree with others of the same type or from a different instrument in the vicinity?
- Hydrostatic checks. This is a more sophisticated checking method applied to data from radiosondes. The thickness (i.e. depth) of standard layers of the atmosphere can be calculated both from the direct measurements of pressure and height, and from the temperature data reported by the radiosonde, using the hydrostatic approximation. If the departure between the two calculations is too large the observation is rejected.

In many operational data assimilation schemes, the flagging of observations is more sophisticated than simply accepting or rejecting the observations. Each observation is assigned a level code which indicates how much weight it should be given during the assimilation. This will depend partly on the type of observation and partly on the results of the checks applied to it prior to assimilation.

4.3.4   The 4D-VAR assimilation process

Once the observations have been processed and quality-controlled, they can be fed into the data assimilation process. As has already been implied, the process is much more complex than simply replacing model predicted values with observational values. This would cause 'shocks' to the model atmosphere, as observations will almost certainly be out of balance with the model atmosphere in a way that would generate high amplitude gravity waves. Instead, the observations are used to iteratively pull the model trajectory away from its initial state towards a trajectory which is more consistent with the observations.

The *data assimilation window* is typically a 3–6 hour period, straddling the specified start time of a particular NWP forecast. For instance, a forecast with a nominal start time of 00:00 UTC may have an assimilation window that runs from 21:00 UTC to 03:00 UTC. All the observations made during this period may be included in the assimilation process as long as they have reached the meteorological centre by the time the data assimilation process starts running. The real time (as opposed to model time) after which no observations are accepted for use in the assimilation is called the *data cut-off time*. A model with a data assimilation window that ends at 03:00 UTC may have a data cut-off time of 03:40 UTC. After this time the assimilation process is started on the computer and no more observations will be accepted. In theory, the more observations used in the assimilation, the more accurate will be the specification of the initial state for the forecast. So a later data cut-off time implies more accurate forecasts. This has to be balanced against the requirement from customers of receiving the forecast as soon as possible in order to be able to use it to inform their decision making. Some forecast centres use longer data assimilation windows or later data cut-off times to improve the performance of their forecast models. This is particularly the case at centres running models for longer periods. In these cases the urgency of getting the forecast out to customers is not so great and a delay of a few hours in issuing the forecast is compensated for by more accurate forecasts at the longer time ranges. For instance, ECMWF, a centre concentrating on issuing forecasts for the 10-day period, uses a 12-hour data assimilation window.

The model used in the data assimilation process is usually a *simplified* and *lower resolution* version of the full NWP model that is used to produce the forecast. Moist processes, such as convection and large scale cloud formation in particular, tend to be very much simplified. Non-linear terms in the dynamical equations also tend to be neglected. This simplified version of the model (called the *forward model*) is run through the data assimilation window period. Observations are inserted into the assimilation at their validity time and, depending on their weighting and error statistics, the model trajectory through the window is adjusted towards the observations. The model also runs *backwards* through the assimilation window in order that observations are able to affect the model trajectory at times prior to their actual validity time. The backward-running version of the model is called the *adjoint*. This forward and backward process happens not once but many times, with the model trajectory being further conditioned by the observations on each iteration. This iteration process continues until the additional assimilation increments become smaller than some pre-determined limit. Typically a model may be run in the order of 50 times through the data assimilation window before this limit is reached. This is the main reason that the model used in the assimilation is run at lower resolution than the actual forecast model, since running at full resolution would greatly increase the time taken to perform the assimilation process.

Figure 4.11 shows a schematic example of the data assimilation process at work. On the horizontal axis of the graph is time through the data assimilation. The vertical axis could be taken to represent several different things. An easy way to consider the diagram is to allow this axis to represent one specific model variable (say temperature) at one specific grid point in the model. A more sophisticated way of looking at the figure is to let the vertical axis be a vector which represents the entire state of the atmosphere across the whole globe. The initial model trajectory through the data assimilation window from the previous forecast is shown in the pale blue line. Observations are shown by green stars. The corrected forecast, a result of the iterative insertion of the observations into the linearized assimilation model as it runs

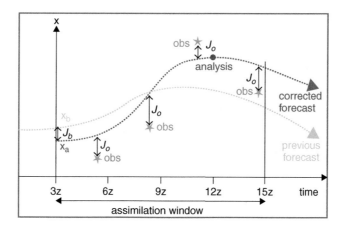

Figure 4.11 A schematic representation of the 4D-VAR assimilation process. The pale blue line shows the model trajectory through the assimilation window prior to the insertion of the observations. The red line shows the adjusted trajectory after the observations (green stars) have been taken into account

forwards and backwards through the assimilation window, is shown by the red line. Note that the corrected forecast is pulled much further away from its original trajectory by some observations than by others. This will depend on the weighting given to each observation. The new trajectory through the assimilation window is an optimum fit to the observations. It is important to realize that, for most operational global forecast models, well over a million observations are used each time the assimilation is run, making the process more complex than shown by this schematic figure.

It is informative to look at the impact of a single observation in a data assimilation scheme. Consider an observation of pressure at around the 400 hPa level over the centre of the Atlantic in a position shown by 'X' in Figure 4.12a. This observation results in a 0.2 hPa increase to the pressure over the initial model background value. This increment is applied in full at the point at which the observation was made but also has an influence on the pressure at surrounding grid points over a pre-determined *radius of influence*. Figure 4.12b shows this for a typical assimilation scheme with a radius of influence of about 1800 km. The influence radius will vary depending on observation type. An observation of a smoothly varying field, such as pressure, will have a much larger radius of influence than a field that varies more sharply over shorter distances, such as humidity. The observation will not just influence surrounding grid points in the horizontal direction but also in the vertical direction. Figure 4.12c shows the increment to pressure in a vertical cross-section. Although the observation was made about 7 km above the surface, it still has a small influence on the surface pressure. However, note that its influence on the levels above it is more limited in

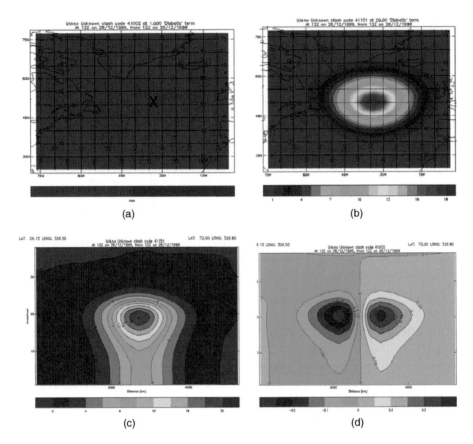

Figure 4.12 Example of the impact of a single observation within a data assimilation scheme: (a) location of a 0.2 hPa increment to pressure at the 400 hPa level; (b) increment added to model background due to that observation at the 400 hPa pressure level; (c) vertical cross-section of increment to pressure along 34°W; (d) vertical cross-section along 34°W of the corresponding wind increments

the vertical. If the pressure field is adjusted during the assimilation process, it is important that all other meteorological variables are also adjusted to be consistent with the change. A change in pressure implies a change in pressure gradient; this in turn implies a change in winds. This is where the data assimilation process really shows its value, since the model will be able to calculate an appropriate wind increment to fit with the pressure increment implied by the observations. This means that all the meteorological variables in the model remain physically consistent with each other during the assimilation process, producing an initial state for the forecast which is balanced. Figure 4.12d shows a vertical section of the wind increments which accompany the pressure increment.

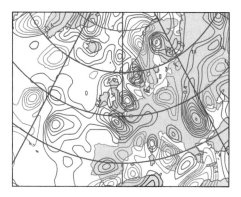

Figure 4.13   Increments to 10 m temperature from the first (left) and second (right) iterations of an operational 4D-VAR data assimilation scheme. Contour interval is 0.2K. Red contours show positive increments, blue contours are negative

Of course, during a real data assimilation run many thousands of observations are being incorporated at once, so the increment patterns generated are far more complicated than the single observation example shown above. Figure 4.13 shows the increments to 10 m temperature from an operational 4D-VAR assimilation scheme.

4.3.5   Information content in NWP analyses

As shown in Chapter 3, the amount of remotely sensed observational data, particularly from satellite platforms, has increased considerably over the past few decades. Much of this additional information is being fed into NWP model assimilation schemes and it is interesting to see what impact it has on the quality of the analysis and subsequent forecast. A study by Cardinali *et al.* (2004) using the ECMWF forecast model attempted to quantify the influence of different observation types on the analysis produced by the 4D-VAR assimilation scheme through the boreal Spring of 2003. The methodology for this study was based on systematically leaving observation types out of the assimilation and assessing the impact on the resulting analysis. The headline finding was that, averaged over the period of the study, 85% of the information contained in the model analysis came from the model background field – that is it was already there in the previous short-range forecast from the model. This is a largely reassuring conclusion since it indicates that, on average, in a short period, the forecast model does not deviate too far from reality. However, it might also lead to the conclusion that the many hundreds of millions of pounds spent on developing, launching and maintaining satellite observing platforms and other observational networks only leads to a 15% impact on the content of NWP analyses. Of course, this 15% is actually the crucial 15%, pushing the model background back towards reality. Particularly in the case of rapidly developing weather systems, the influence of observations is critically important in setting the initial conditions for the weather forecast.

Of the 15% of the information in the analyses that came from observations, about 25% was found to come from the various *in situ* platforms (land-based and ocean SYNOP observations, drifting buoys, radiosonde soundings, aircraft wind and temperature reports and wind profiler measurements) whereas the remaining 75% came from the various satellite-borne instruments (wind estimates from scatterometers and cloud drift vectors, infrared and microwave radiances etc.). Here the influence of the expensive satellite-based observations now becomes clear. Since 2003, ECMWF has continued to increase the amount of satellite-based observations that it uses in its assimilation scheme (Figure 3.15), so one might expect the proportion of information within the analysis coming from satellites to have increased accordingly. Satellite data have the largest influence on the initial conditions, and hence the subsequent forecast accuracy, in the Southern Hemisphere, where ground-based observing systems are sparser than in the Northern Hemisphere. Tests at ECMWF have shown that the impact of *not* using satellite data in the data assimilation for their global model is that the forecast in the Northern Hemisphere loses useful skill less than a day earlier than if the satellite data are included. By contrast, in the Southern Hemisphere the forecast loses useful skill about 2.5 days earlier if the satellite data are not used (Kelly and Thépaut, 2007b).

Some interesting characteristics can be discerned by looking at the influence of the different observation types separately. For instance, surface pressure observations from drifting buoys have a very low influence on analyses when considered globally due to the low number of observations. However, their influence *per observation* is very high as they are in otherwise data sparse regions and often provide the first signs of rapid pressure changes as the precursor to the development of an active weather system.

4.3.6   *Assimilation of satellite data*

Because of the importance of satellite data in setting the initial conditions for the forecasts, particularly over otherwise data-sparse regions (Section 4.3.5) it is worth spending some time considering how satellite data are incorporated into the assimilation cycle. This is an extremely complex process. It is important to realise that, although satellite observations are used to set the initial conditions of atmospheric temperature and humidity, satellites *do not measure these quantities directly*. The only quantity measured by satellite instruments is the intensity of radiation in different wavelengths of the electromagnetic spectrum – that is *radiances*. The radiance in a particular wavelength measured at a satellite is related to the temperature of the object that emitted that radiation, but it will also be related to other factors, such as how much of that radiation was absorbed or scattered between the emitting body and the satellite. Satellite humidity measurements are even more complex. It is well known that water vapour absorbs and emits radiation very strongly in particular parts of the infrared and microwave sections of the electromagnetic spectrum, so satellite instruments are designed to measure radiation at those particular wavelengths. However, the emission and absorption of radiation by water vapour is a function of its temperature as well as its mixing ratio, so inferring water vapour mixing ratios from satellite radiances is a very tricky problem.

All satellite-based instruments record radiances at multiple wavelengths – referred to as 'channels' – in order to sample radiation emitted in different parts of the atmosphere. Because, to the first order, the temperature of the atmosphere is a function of height above the Earth's surface, the different channels can be taken to be representing the radiation emitted at different levels in the atmosphere. This allows a set of *weighting functions* to be defined which indicate which layer of the atmosphere the radiation in each channel has been emitted by. Figure 4.14 shows a set of weighting functions for the Advanced Microwave Sounding Unit A (AMSU-A) instrument on board the NOAA polar orbiting satellites which are used for temperature retrievals.

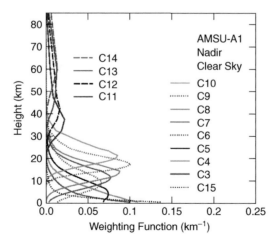

Figure 4.14 Weighting functions for the 13 channels of the Advanced Microwave Sounding Unit (AMSU) satellite-borne instrument which are used for temperature retrievals

Thirteen channels are used which allow temperature to be retrieved from about 3 hPa down to the surface. Because of the overlapping nature of the weighting functions, together with the complex absorption, scattering and emission of radiation, there is no one unique temperature profile which would account for the radiation received at the satellite in each of these channels. In order to specify a temperature profile given a set of satellite radiances it is necessary to have *a first-guess profile* of temperature which these radiances can then be used to adjust to fit the observations. It is also clear that satellite radiances will not be able to resolve very fine structure in the temperature profile, such as boundary layer inversions or inversions at frontal surfaces or the freezing layer.

Because of the need for a first guess of the temperature profile associated with a particular set of radiances, together with the complex issues of emission, scattering and absorption of radiation, many forecasting centres use the data assimilation process to deal with the satellite radiances rather than converting the radiances into temperature and humidity values prior to assimilation. The existence of a first guess of the state of the atmosphere is an integral part of the data assimilation process and the model's radiation scheme can be used to compute the various emission, scattering and absorption processes that lead to the particular radiance values measured at the satellite. The presence of other observations, such as balloon soundings and aircraft measurements, will also be able to adjust the temperature and humidity profile in accordance with the known error characteristics of each type of instrument.

4.3.7 'Intervention' and 'bogussing' in the assimilation process

As stated previously, data assimilation schemes are generally set up to be more likely to reject or down-weight observations which are a long way from the background trajectory in order to minimise occasions when 'shocks', which can lead to instabilities, are introduced into the model. There have been a number of cases where the inherent caution of a data assimilation scheme has led to the rejection of correct observations which indicated the presence of a rapidly developing weather system that went on to cause significant damage. This is particularly the case over data sparse regions such as oceans. A few forecast centres attempt to reduce the chances of this type of situation occurring by allowing a human forecaster to monitor the status of the observations as they are processed by the assimilation scheme. Observations of, for instance, rapidly falling pressure may be rejected by the automatic processing but comparison with available satellite imagery may show the early signs of a rapidly deepening weather system in terms of a baroclinic cloud leaf structure or an emerging cloud head.

Having identified that the observations are likely to be correct, the forecaster can then 'support' those observations. The usual way to do this is to invent lots of *bogus* observations in the vicinity which all indicate the same pattern of falling pressure. Faced with a large number of observations all saying the same thing, the assimilation scheme is forced towards the correct observation, although due to the constraints built into the code, it is still unlikely to go all the way towards the correct solution. Figure 4.15 shows the UK Met

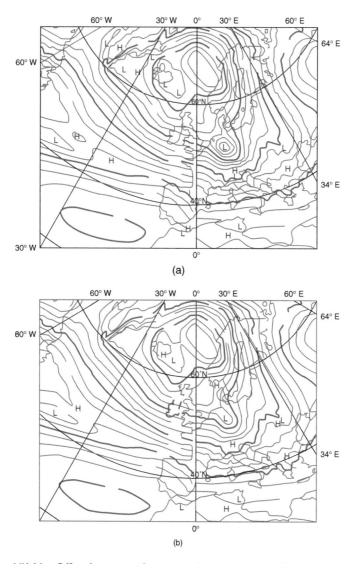

Figure 4.15   UK Met Office forecasts of mean sea level pressure on 26 December 1999 (a) with and (b) without forecaster intervention to support ship observations of rapidly falling pressure in the Atlantic 24 hours earlier. (© Crown Copyright 1999, Met Office.)

Office T+24 forecast of the Lothar storm of December 1999 both with and without forecaster intervention to support ship observations in the Atlantic 24 hours earlier. The only difference between the two pictures is the depth of the depression over north-west Europe. The forecast with intervention has produced a much deeper low resulting in a much bigger pressure gradient and, hence, stronger wind across northern France and western Germany. (See Figure 2.1b for the verifying analysis for this storm.)

The process of human intervention relies heavily on the ability of a forecaster to spot the signs of rapidly developing systems and then to be brave enough to effectively bet against the model by supporting the appropriate observations with well formulated bogus ones. The UK Met Office also runs an *automated* bogussing system for developing tropical storms. Due to their nature, tropical storms form over tropical ocean regions which are rather data sparse, and the presence of a shield of thick cirrus cloud usually limits the amount of available observational data from satellite retrievals. Ships and aircraft will also try to avoid them. The UK Met Office bogussing scheme uses the tropical storm reports from the major storm warning centres which include estimates of wind strengths and patterns within tropical storms to produce a set of bogus aircraft reports of wind speeds around the system. These can be inserted into the assimilation process to produce more accurate representations of the storms in their initial phases of development, which then leads to more accurate forecasts of these systems.

4.3.8   *Future developments in observing systems and data assimilation*

The increasing amount of remotely sensed and, in particular, satellite-based observations has led to the development of a number of different statistical tools for assessing the impact of different observations on the quality of the forecast. Further development of these tools will lead to more optimal strategies for deciding which observations to use and which new observation types or platforms are likely to provide value for money in terms of improvement to forecast skill.

One particularly exciting development currently under consideration is the use of an *interactive forecast system*. In such a system, the data assimilation scheme would be able to assess where extra observations would have the maximum possible impact on the quality of the forecast over a particular region of interest. The interactive system would then launch additional observations in this *optimal zone for observing* (OZO), possibly by firing off rocketsondes from moored buoys or by launching instrumented unmanned drones. This additional information could then be incorporated into a new run of the assimilation scheme. There is clearly an overhead here in the extra time that it would take to collect the new observational data, plus of course the financial cost of setting up and maintaining such a system, but this may be thought worthwhile in particularly awkward forecasting situations where a forecast outcome was very heavily dependent on small changes to the initial conditions and where the extra observations would greatly increase the confidence in the forecast.

Summary

- NWP models are designed to simulate all the physical processes considered important to determining how the atmosphere will evolve in time.
- Whilst some physical processes can be represented explicitly, many others have to be dealt with through parametrization.
- The amount of physical detail that can be included in NWP model formulation is limited by both our scientific knowledge of the physical processes themselves and by available computing power.
- Data assimilation is a crucial part of the NWP process, blending observations that contain information about the state of the atmosphere with a numerical model that embodies the laws of physics which determine how that state will evolve.
- Observational data from satellites dominates the information content in the initial conditions of a weather forecast.

5

Designing Operational NWP Systems

Chapter 4 described the various ingredients of an NWP computer model. Any operational NWP system will need to include these elements but there are many different ways in which they can be combined and implemented. In this chapter we shall consider the kinds of decisions that major forecasting centres have to make in order to design and build forecasting systems that meet their needs and the demands of their customers. These decisions will be determined by a wide range of different factors, not least of which is the availability of funding to pay for high performance computing facilities and staff to develop, maintain, run and analyse the NWP systems.

5.1 Practical considerations for an NWP suite

The major determining factor on the design of any operational weather forecasting system is the availability of computing power and funding for the staff to develop and program computer models, maintain and operate the various systems and analyse the output from these models. Most major operational centres have long-term plans which are predicated on the increases in computing power expected over the coming decade or so, with upgrades to forecasting systems planned to coincide with the acquisition of more powerful computing facilities. The rate of progress in developing more powerful computers is accelerating and this is one reason why this chapter may become out-of-date rather more quickly than other parts of this book.

Given a certain amount of computing power, operational forecasting centres need to make many decisions on how they will design their forecasting systems to make optimal use of that power. These decisions will be partly informed by the requirements of the customers of the NWP forecasts. Here we are using customers in a very wide sense, which does not just include

Operational Weather Forecasting, First Edition. Peter Inness and Steve Dorling.
© 2013 John Wiley & Sons, Ltd. Published 2013 by John Wiley & Sons, Ltd.

commercial customers who pay for forecast information. The customers of an NWP system include the forecasters who will use the NWP output to produce forecast information. Some of these forecasters may be working within the same organisation as the operational centre or may themselves be employed in other organisations providing forecast information to third parties. End users of weather forecasts include the military, commercial organisations such as airlines, shipping lines and energy suppliers, the general public via media such as TV, radio and internet providers and government organisations who may have to deal with the effects of extreme weather events such as environmental protection agencies. All of these customers will have a set of requirements for the forecast information they receive, and the design of an NWP system must try to take into account as many of these requirements as possible in order to provide a useful service to its customers.

Customer requirements of forecasts fall into a number of different categories:

- **Geographical coverage**. Are forecasts required over a wide area (possibly global or continental) or over a much more geographically limited region? This geographical coverage also extends to coverage in the vertical. Commercial airlines, for instance, may require forecasts of winds at many different levels in the atmosphere.
- **Geographical resolution**. Are forecasts required that can distinguish small scale weather variations over short distances, such as the temperature difference between an urban and surrounding rural area, or is a more broad brush approach acceptable?
- **Level of detail**. Do customers need to have detailed quantitative forecasts of many different weather parameters or is a simple forecast of perhaps only one variable at a single location sufficient?
- **Forecast range**. How far ahead do customers want weather forecasts? As has already been discussed in Chapter 2, deterministic forecasts are generally not practical beyond about 14 days, but if all that customers want are forecasts out to three days ahead it makes little sense to implement a system which produces forecasts out to the deterministic limit.
- **Specialised information**. Are there customers who require specialised information – for example, detailed information on ocean wave heights and periods or the ability to track atmospheric dispersion of pollutants? These sorts of forecasts may require additional, specialised models which are designed to forecast these specific quantities, often using the output from a weather forecasting model as input.

These requirements will determine choices on many of the aspects of an NWP system, which were outlined in Chapter 4. The domain of the model (in both the horizontal and vertical), the level of detail included in the model physics, the grid box size and the nature and level of detail of the data assimilation system will all be affected in some way by this list of requirements. It may well be that for a major forecasting centre, such as a National Meteorological Service, the list of requirements from customers is so diverse that more than one NWP model has to be used to fulfil them all. This results in what is known as an 'NWP suite', and the way that the different models within such a suite interact with each other leads to another set of decisions that need to be made in order to get maximum effectiveness from the available computing resources.

These considerations will affect not only the design of NWP models but also much of the other apparatus of operational forecasting. What observations need to be made in order to feed the NWP system with appropriate initial information? What level of further processing does the output from the NWP system need to be subjected to (either by automated systems or a human forecaster) in order to produce appropriate forecasts for customers? How many human forecasters need to be involved in the process of forecast production for customers? What platforms are the most appropriate for the delivery of those forecasts?

5.1.1   Model domain

Probably the first consideration to be made when designing an NWP model is the domain – that is the geographical coverage. In this respect, numerical models divide into two broad categories – global models covering the whole globe and regional models covering a more limited area.

Global models are, in some senses, a foundation of any NWP suite. Even if forecasts are only required over a limited area, such as an individual country, numerical models running over such a limited area still need to know about the weather systems outside that region. The Earth's atmosphere is a continuous fluid and weather events over one part of the globe can potentially affect the development of the weather anywhere else over the globe, particularly as forecast lead time increases. So an NWP model that only covers a small region such as an individual country still needs to be fed information about the weather systems arriving at the boundaries of its limited domain. This information will most likely come from an NWP model with a global domain, so for most major national meteorological services a global domain NWP model is the basis of their forecasting system. Many of these national meteorological services will have some customers who require global information anyway. Long haul airlines and global shipping companies are two obvious customers for which global forecast coverage is a requirement.

Benefits of a global model:

- The most obvious benefit of a global model is the global coverage. Wherever a customer requires forecast information for, a global model can provide it, although this will be at the expense of high spatial resolution.
- As stated above, the atmosphere is a continuous fluid with no edges. A global model, therefore, does not need to deal with the edge issues associated with limited domain models. Global models are self-contained and do not need to be fed with information from outside their domain. Edges in a numerical model are also potential sources of numerical instability, so not having any edges avoids a range of potential problems involved with dealing with artificial edges and calculating gradients at boundaries.
- As forecast range increases, the weather at any point on the globe becomes more likely to be affected by weather at increasingly distant locations. Hence, to make forecasts at longer lead times, a global model is necessary.

Drawbacks of global models:

- As always with NWP, the ultimate restriction on model design is available computing power. Producing numerical forecasts over the entire globe means that the resolution of the model needs to be restricted in order to perform the forecasts within a reasonable length of time. So global models produce global coverage but at the expense of geographical resolution.
- Although global models remove the issue of artificial horizontal edges, they still have numerical issues when dealing with flow near to or over the poles. If the earth's surface is rendered onto a regular latitude–longitude grid, the grid boxes will become smaller in the longitude direction near to the poles, so it becomes more likely that the CFL condition discussed in Chapter 4 may be violated. The pole itself becomes a singularity in an otherwise regular grid. Various numerical methods have been devised to overcome these problems.
- Depending on the actual grid size of the global model, many physical processes which occur on scales smaller than the grid length will need to be parametrized.

Having chosen a global domain, the next decision is where to put the top of the model. Global models tend to have deeper vertical domains than regional models. This is partly because of the fact that global models tend to run out to longer forecast lead times. It has already been shown that at longer forecast lead times, the weather at a particular location is more likely to be affected by the weather at increasingly distant locations. This is equally true in the vertical. Disturbances in the stratosphere, such as sudden changes in the strength of the polar vortex over the winter hemisphere, can propagate downwards and affect the state of the troposphere over periods of days to weeks. To capture these effects a global model needs a well represented and resolved stratosphere. The current generation of global NWP models typically has a model top well into the upper stratosphere or even mesosphere at around 80–90 km above the surface.

Figure 5.1 shows the vertical levels in the UK Met Office global NWP model prior to 2006 on the left and between 2006 and 2009 on the right. Below about 15 km the two versions were identical but above this height the 2006–2009 version had approximately twice as many levels in the lower to mid-stratosphere and extended up to a height of 65 km, which is in the mesosphere. The pre-2006 version had its top at 38 km, so did not include the upper stratosphere. In 2009, the UK Met Office upgraded its model again, taking the model top up to 80 km.

Another consideration for the vertical domain of a global model is based on atmospheric chemistry. The stratosphere is a very active region of the

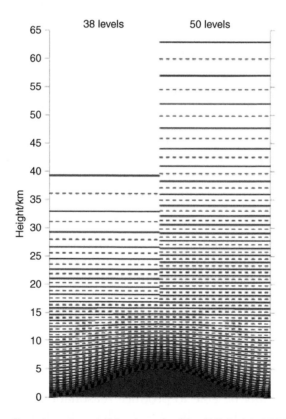

Figure 5.1   The configuration of vertical levels in the Met Office global NWP model. (Pre-2006 version on the left and 2006–2009 version on the right.) Black lines show the model main levels and dashed red lines show the 'half-levels' on which some variables are calculated and stored. (© Crown Copyright 2009, Met Office.)

atmosphere in chemical terms, with the constant destruction and renewal of ozone occurring in the mid-stratosphere. Some specialised customers of weather forecasts, such as health protection agencies, would like information about this process, with predictions of ozone concentrations, so this is another reason for extending the model top above the stratosphere. Of course, the interaction of ozone with sunlight is also an important part of the radiation budget of the atmosphere, so explicitly modelling ozone variations through the period of a forecast ought to lead to a better representation of the Earth's energy budget.

Many customers of forecasting centres want very detailed weather forecasts for specific locations. Some customers may even want to be able to distinguish conditions between sites separated by only a few kilometres. To provide these kinds of forecast, forecasting centres will run numerical models with very fine grids, with a grid box size of perhaps only a few kilometres. Due to the limitations of computing power, it is not practical to run such models over very large domains – the forecasts would take too long to produce. Hence, forecasting centres will run *regional* models with a domain that covers just the area of their main customer's interests, with perhaps a small surrounding region included. Typically for a national weather service this domain would be chosen to cover the area of the country concerned, with perhaps an extension in the direction from which most of the important weather systems arrive. Some major centres even run more than one regional area model. The UK Met Office, for instance, runs a model covering a large area of the North Atlantic, as far west as Newfoundland, and western Europe at a moderate resolution (grid boxes 12 km square in 2011) and then a more limited domain model over the United Kingdom and Northern France with a much higher resolution (grid boxes 1.5 km square in 2012).

Benefits of regional models:

- The most obvious benefit is higher resolution. Geographical features such as hills and coastal features that affect the weather will be better represented in higher resolution models, hopefully providing better local detail in weather forecasts.
- In higher resolution models it may be possible to switch off certain parametrization schemes as the processes that they represent will be explicitly resolved on the model grid.
- High resolution models should be able to provide better detail for local forecasts than coarser resolution models. For instance, a model with a 25 km grid box may be able to produce reasonably accurate forecasts of large scale variables, such as the pressure distribution and winds above the boundary layer, but may have considerable errors in forecasts of variables at the surface, such as maximum and minimum temperature forecasts or surface winds for specific locations. This means that the model output has

to undergo considerable further modification, either by a human forecaster or a computerised algorithm, in order to produce accurate forecasts of surface quantities. A high resolution model, which also has good vertical resolution in the boundary layer, may be able to produce forecasts of these types of quantity to a high enough accuracy to remove the need for the post-processing effort. Thus, these types of forecast can be produced much more rapidly, and potentially for many more locations.

Drawbacks of regional models:

- As discussed, the presence of artificial boundaries can cause numerical problems for limited domain NWP models. The boundary regions need to be chosen and treated carefully to avoid numerical modes appearing due to the presence of an artificial edge. In particular, any reflection of weather features by the boundaries must be suppressed. Calculating gradients at a boundary also requires careful handling to avoid loss of accuracy.
- All regional models need to be fed at the lateral boundaries to give information about weather systems propagating into the domain from outside. Thus, no regional model can be run as a stand-alone forecasting tool and a global model must also be available to provide the information about what is happening outside the regional model domain. The global model providing this information is usually referred to as the 'driving' model.
- The issue of parametrization becomes rather complicated at certain grid box sizes. For instance, with grid boxes 5 km square, most convective clouds are still smaller than the grid scale and require parametrization, but large convective cells may be resolved by the grid. Thus, is may not be clear whether a convective parametrization needs to be used or not.

Choosing the optimum domain and resolution for a regional model involves making some careful choices. Ultimately, a forecasting centre would like as large a domain as possible, with the highest possible resolution. The logical conclusion is that a high resolution global model would be the preferred option but clearly issues of computing power rule this option out. There are several possible compromise solutions.

The simplest solution is to cover the area of interest with a high resolution grid and then extend the boundaries as far away as possible from the area of interest whilst staying within the limits of available computing power. Ideally, the boundaries should be extended furthest in the direction from which the prevailing weather conditions arrive most often. The aim here is to try to achieve a domain such that all the weather that affects the area of interest develops within the model domain during the period of the forecast. Clearly, this will not always be possible in fast moving weather situations.

In such cases the information being fed in at the boundaries from the global NWP model becomes critically important in determining the accuracy of the forecast.

A second possible solution is to run a set of models which 'nest' within each other. A moderate resolution regional model could be used to cover a wide area, with a higher resolution regional model sitting within this larger model, positioned over the area of greatest interest. The model with the larger domain would feed information to the boundaries of the higher resolution model. An obvious drawback here is that two models need to be run, with implications for the use of computing resources and the time taken to produce the forecasts. It is not the case that the first model needs to complete its forecast before the second one starts running. If a sufficiently large computer is available, both models can run at the same time, as long as the larger domain model stays slightly ahead of the smaller one. Boundary data can then be passed from the larger to the smaller as the models run, thus reducing the total integration time (Box 5.1).

Box 5.1 An example of boundary choices in regional models

The UK Met Office runs two regional models. The first covers the entire North Atlantic, all of Europe and a small section of North Africa (outer box in Figure B5.1.1). The second covers just the British Isles and the nearest parts of continental Europe (inner box in Figure B5.1.1). The larger of these two configurations runs forecasts out to 48 hours and the smaller out to 36 hours.

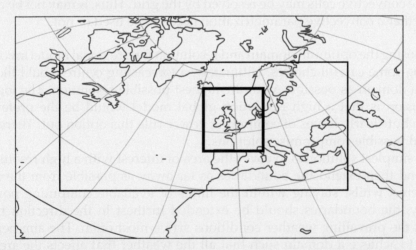

Figure B5.1.1   The regional model domains used by the UK Met Office. (© Crown Copyright 2009, Met Office.)

The larger of the domain is extended westwards from Europe out into the Atlantic. This domain is designed to capture the development of the Atlantic depressions that constitute the major weather systems affecting western Europe. The majority of these will form and develop within the domain on a 48-hour period. This domain is referred to as the North Atlantic European (NAE) model.

The smaller domain is extended to the south and east of the United Kingdom. This domain is designed partly to optimise the ability to capture the vigorous convective events which form over Continental Europe and which sometimes propagate over the United Kingdom during the Summer months. Again, on the 36-hour timescale of this model, all the development should take place within the domain. This model also has a smaller grid box size in order to better resolve organised convective systems

There will, of course, be occasions when the weather that affects the United Kingdom during the forecast period actually starts developing outside the domains of these models and then the forecast accuracy becomes more dependent on the data being fed in at the boundaries from the global NWP model.

Another solution to the domain versus resolution problem is to use *variable resolution* over a larger domain. This involves using a horizontal grid which does not have uniform grid spacing. Instead, the grid spacing is very fine over the area of greatest interest but then increases towards the boundaries, where detailed resolution is not so important. This is particularly effective if the boundaries are over oceanic regions where there are no detailed topographical features which would benefit from being highly resolved. Another operational use of this type of variable resolution model is in tropical storm forecasting. A regional model can be defined over a wide area of potential storm activity, with the highest resolution defined over the area of greatest interest for storm track and intensity forecasting. Models are also being developed and tested in which the region with highest resolution is able to track along with a particular weather feature, such as a tropical storm, meaning that the highest resolution is always in the area where it is most beneficial.

Dealing with the boundaries is the most problematic aspect of using regional models for NWP. Information needs to be fed into the domain from a global (or larger domain regional) model and clearly these data will be on a different grid to the high resolution model. Hence, interpolation is required. Potentially even more problematic is that there may be a meteorological mismatch between the two models, with unrealistically large gradients or even discontinuities being possible across the boundaries. This problem is overcome by using a 'buffer zone' around the edges of the regional model.

In this buffer zone, the information on the outer ring of grid points will come entirely from the global model. At the next ring in from the boundary, the information will be a weighted average of the global and regional model fields, with the weighting being biased towards the global model (perhaps 75% global and 25% regional). The weighting given to information from the global model is further reduced at the next ring of points into the domain, and reduced once again at the next ring until all the information is being calculated by the regional model on the inner edge of the buffer zone. This configuration prevents mismatches between the two models which might cause instabilities to develop through the spurious occurrence of unrealistically large gradients of quantities such as pressure or temperature. Careful numerical methods need to be applied in the buffer zone to prevent signals which are propagating outwards from the interior of the regional model then reflecting at the boundary and moving back into the interior of the domain.

Further refinements of this buffer zone approach are possible. One obvious approach is to make the conditions in the buffer zone dependent on the direction of movement of air parcels. In cases when the winds are blowing into the domain the winds are taken from the global model. When the winds are blowing out of the domain, the advecting winds are extrapolated from the interior of the regional model.

The quality of forecasts from a regional model is highly dependent on the information coming from the global driving model. In a situation in which, for instance, a rapidly moving frontal system is due to propagate into the regional model domain during the forecast period, its timing will be largely determined by the global model. It cannot arrive within the regional model domain without having been passed in from the global model through the boundary. Thus, if the timing in the global model is wrong, it will also be wrong in the regional model. The intensity of such a system will also be largely determined by its handling by the driving model. Although once the system is within the regional model domain, the higher resolution model may be able to modify its intensity, this will take a finite time and will anyway be partly determined by information propagating in at the boundary.

5.1.2   *Model resolution*

Going hand-in-hand with choice of NWP regional model domains is the question of resolution. Customer requirements on the level of local detail will be important considerations here, as always limited by the constraint of available computing power. There is an obvious argument that higher resolution will lead to better forecasts, as topographical features which affect local weather conditions are better resolved and numerical errors introduced by discretising the equations of atmospheric motion onto a numerical grid are reduced. However, constant striving to increase the resolution of numerical models could be seen as simply throwing money at the forecasting problem unless it is also accompanied by development of both the numerical methods used to solve the equations and the physical parametrizations used to represent subgrid scale processes. Many aspects of NWP model parametrizations turn out to be resolution dependent and are tuned to produce the best forecasts when applied within a model with a particular grid spacing. A simple increase in resolution without further testing and tuning of the model physics can actually lead to a *deterioration* of the quality of the forecasts. In some cases simply tuning parameters within the physics schemes may not be enough to correct for this and a completely new scheme more appropriate for a higher resolution may need to be implemented. In some cases, as resolution increases, some physical parametrizations can even be switched off as the process they are designed to represent actually becomes resolved on the model grid.

In general, if carefully implemented, increases in resolution lead to better forecasts (Box 5.2). The intensity of synoptic scale systems is often better represented in higher resolution forecasts, and local weather detail is usually better captured in models in which the local geographical features are better resolved too. Figure 5.2 shows two forecasts of Hurricane Katrina close to its peak intensity in August 2005. Both are from the ECMWF global model, one with T511 spectral resolution (which was the operational resolution of this model at the time) and one with T799. At first glance the forecasts appear similar. However, the higher resolution version has much more intense precipitation and the central pressure of the hurricane is 31 hPa deeper – 909 hPa compared to 940 hPa at lower resolution. In fact, the actual central pressure at this time was about 920 hPa, so the high resolution version has somewhat over-deepened the system.

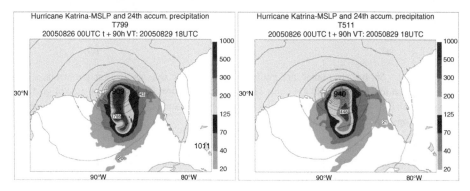

Figure 5.2   Two forecasts of Hurricane Katrina from the ECMWF global model. Left panel – T799 spectral truncation. Right panel – T511 spectral truncation. T511 is about equivalent to a grid spacing of 40 km, whereas T799 is about equivalent to a 25-km grid. Contours are sea level pressure and colours are precipitation accumulation in 24 hours. (Reproduced by permission of ECMWF.)

Box 5.2 Benefits of high resolution – an example

On 16 August 2004 a small area on the north coast of Cornwall in the United Kingdom experienced a series of convective storms which brought over 200 mm of rainfall to the catchment of the River Valency. This caused flooding in the village of Boscastle at the river mouth, destroying buildings and bridges. The flooding was so abrupt and intense that many people had to be airlifted from the roofs of buildings by helicopter. A trigger for these storms was convergence of low level winds caused by the configuration of the coastline of North Cornwall.

At the time the highest resolution NWP model run by the UK Met Office had grid boxes 12-km square. This model completely failed to capture the intensity of the rain in the region. Tests of a model with 4-km square grid boxes, which was in development at the time, showed that the higher resolution model was capable of capturing the intensity of the rain although the area of heaviest rainfall was slightly displaced in this forecast (Figure B5.2.1).

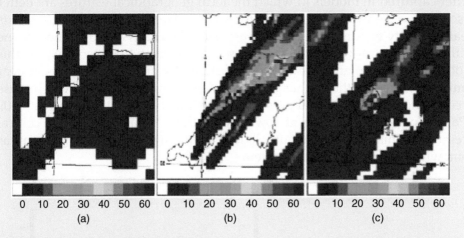

0	10	20	30	40	50	60	0	10	20	30	40	50	60	0	10	20	30	40	50	60
			(a)							(b)							(c)			

Figure B5.2.1   Six-hourly rainfall accumulations (mm) over the south-west peninsula of the United Kingdom for 12:00–18:00 UTC on 16 August 2004 from: (a) Met Office 12-km grid box regional model: (b) Met Office 4-km grid box regional model; and (c) weather radar observations. (© Crown Copyright 2004, Met Office.).

The 4-km grid box model gives a clear indication of the potential for very heavy rain and flooding in the North Cornwall area and would have provided good guidance for the issue of heavy rain warnings. Although the heaviest rain predicted by this model is not quite in the right place it is clear that the convergence line has been resolved by the higher resolution model.

It is important to realise that, even in NWP models with high resolution grids, many geographical features which may affect local weather conditions will still only be partially resolved at best and often go unresolved completely. Even when moving to grids of the order of 1 km, narrow mountain ridges, small valleys and coastal bays and headlands may go unresolved on the grid. Local sources of water, such as small lakes and reservoirs, may also not be represented. All of these features can cause local variations in weather conditions, and hence there is still plenty of scope for NWP forecasts to be improved by the presence of a human forecaster with a good knowledge of the local weather conditions within a forecast area.

5.1.3   Designing and running an NWP suite

The preceding discussion makes it fairly clear that a major forecasting centre will best meet the needs of its various customers through the use of more than one NWP model. The actual composition of this suite of models will be determined by several factors. The over-riding factor as always will be availability of computing resources, but another factor will be the total area for which the forecasting centre has a responsibility to produce forecasts. Both the National Weather Service in the USA and the Met Office in the United Kingdom are designated World Area Forecast Centres (WAFCs) for the production of weather forecasts for global civil aviation. For this and other reasons both forecasting centres run a global domain forecast model. Both forecast centres have a wide range of customers which require detailed location-specific weather forecasts, the majority of whom want information for sites within the national boundaries of their countries. However, with the United Kingdom covering only about 243 000 square kilometres, an area smaller than the single US State of California (423 000 square kilometres), it is clear that the provision of high resolution weather forecast information by the two national meteorological services will be quite different.

The models which comprise the forecast suite of the UK Met Office have already been discussed, but are listed again here:

- A global model, which at the time of writing had a grid spacing of approximately 25 km, 70 levels in the vertical with a model top at approximately 80 km above the surface. This model produces forecasts out to six days ahead.
- A regional model covering most of the North Atlantic and all of Europe (the NAE domain). Grid spacing of approximately 12 km with 70 levels in the vertical and a model top at about 80 km. Forecasts are run out to 48 hours ahead.

- A regional model covering the United Kingdom, parts of continental western Europe and surrounding waters (the UK4 domain). Grid spacing of 4 km with 70 levels in the vertical and a model top at about 40 km. Forecasts are run out to 36 hours ahead.
- In addition to the single run of the global model out to six days ahead, the UK Met Office also runs a global model 25 member ensemble out to three days ahead. This ensemble runs at a lower resolution with a grid spacing of approximately 60 km.

Figure B5.1.1 shows a map of the regional model domains.

It is important to optimise the forecasting system by linking these models together in a way that produces the required forecast information efficiently. Which model should run first? How often should each model be run and how far into the future? Which set of boundary conditions should be used to provide driving data for the regional models?

To illustrate the process of linking a suite of NWP models together into an operational schedule, Table 5.1 shows the sequence of model runs that comprise the UK Met Office operational suite. The table just shows the morning part of the schedule. The schedule repeats itself in the afternoon.

The first point to note here is that the global model is not the first model to be run in the suite. Instead the regional model covering the North Atlantic and Europe is the first in the sequence. This model produces a reasonably detailed forecast (grid spacing about 12 km at time of writing) over a broad region. Hence, this model will satisfy a lot of customer's requirements, providing as it does a reasonably broad coverage and moderately high resolution. A potential downside to this arrangement is that this regional model has to be driven at the boundaries with information taken from an earlier run of the global model – the 18:00 UTC run from the day before. Although this run of the global model is six hours out-of-date, it is updated at 21:00 UTC with a run

Table 5.1   The morning schedule for the UK Met Office operational model suite as in 2011. This schedule repeats itself during the afternoon (add 12 hours to all times)

Real time (UTC)	Model name and start time (UTC)	Forecast range (hours)	Lateral boundary conditions provided by
01:30–02:10	NAE (00:00)	48	Global 18:00 UTC
02:40–04:15	Global (00:00)	144	N/A
04:20–04:55	UK4 (03:00)	36	Global 00:00 UTC
05:15–06:55	Global ensemble (00:00)	72	N/A
07:30–08:10	NAE (06:00)	48	Global 00:00 UTC
08:45–09:45	Global (06:00)	48	N/A
09:30–10:10	UK4 (09:00)	36	Global 06:00 UTC
10:15–12:15	Computer maintenance		

of the data assimilation cycle to bring it back towards observations. The small degradation in the forecast due to running from older boundary conditions is more than offset by the ability to get forecasts out to customers much more quickly than if the NAE model run had to wait for a full integration of the global model at 00:00 UTC.

Next to run is the global model, which produces a forecast out to six days (144 hours). Since this is a global model it requires no lateral boundary conditions. This model provides the global coverage and medium-range lead time that are required by a wide range of different forecast users. Following on immediately from the global mode is a run of the high resolution (4 km grid) UK regional model, out to 36 hours ahead, starting from conditions at 03:00 UTC. This regional model is driven at the boundaries by the global model which has just completed, meaning that the boundary conditions are as up-to-date as possible. This model run provides all the local detail needed for site-specific forecasts in the United Kingdom. It is timed to finish running just before 5 a.m., so that its output is available to forecasters who are producing forecasts for customers who are just starting to get into work and will be making decisions on their operations for the coming day at this time of the morning. Forecasts on national and local television and radio will also draw on the output of this model run. The timing of this forecast model run is a good example of how the requirements of customers are of paramount importance when a major national meteorological centre is designing an NWP suite.

Following the run of the UK model, the UK Met Office then runs its global model ensemble. This is a 25-member ensemble run out to 72 hours with a grid box size of approximately 60 km. The control member of the ensemble runs from the same initial conditions as the 00:00 UTC run of the full global model, with the other 24 members having perturbed versions of these initial conditions. This model run is primarily useful to the senior forecasting team who can use it to make an assessment of the uncertainty on the forecast over the next three days. Having an ensemble also allows for the generation of probability forecasts based on the behaviour of the various ensemble members.

The 06:00 UTC run of the North Atlantic and Europe (NAE) regional model is next in the schedule. In a similar way to the 00:00 UTC run of this model, the boundary conditions come from the previous run of the global model. This model run gives a broad regional picture of the forecast for the next 48 hours, allowing forecasters to check the predicted evolution and make updates to forecasts they have already issued if necessary. It is then followed by the 06:00 UTC run of the global model. This run only goes out to a lead time of 48 hours (compared to the six- day lead time of the 00:00 UTC run) and its primary purpose is to provide an up-to-date set of boundary conditions for the 09:00 UTC run of the UK 4 km regional model which follows it in the schedule.

Following a period of computer maintenance, the afternoon schedule follows the same pattern as the morning.

Clearly, this schedule provides a vast amount of forecast information at a variety of different resolutions and lead times. This would not be possible without a seriously powerful computer. In fact, the computers used by National Meteorological Centres are often amongst the most powerful computers in the world and there are very few other applications which make such heavy operational use of high performance computing. Operational forecasting centres upgrade their computers regularly and are usually planning developments in their NWP capability on a 5–10 year timescale, predicated on the availability of increasing computer power. The advent of an operational ensemble forecast in the UK Met Office suite for instance has been planned and developed over several years but was only made possible by a recent computing upgrade. As well as running the operational suite of forecasts, the computer will also be used to test new configurations of the forecast models that are currently under development and for computing forecast verification statistics which give a picture of the recent performance of the operational forecasts. Hence the maintenance of the computing systems and the scheduling of the various tasks is as vital a task within a forecasting operation as the interpretation of the NWP model output and forecast production for customers.

5.2   Ensemble prediction systems

The nature of the weather forecasting problem and the inherent unpredictability of the atmosphere were discussed in detail in Chapter 2. This unpredictability leads to a limit on deterministic weather forecasting which is, at best, about 14 days. However, in some cases the limit may actually be considerably shorter than this. There is no way to know, *a priori*, how predictable the atmosphere is on any given day but it would certainly help forecasters if there was some way of making, at the very least, a subjective assessment of this predictability. This will not necessarily make the forecast any more accurate but it will allow forecasters to convey to their customers how much confidence to place in their forecasts.

In operational forecasting, the normal way of assessing the predictability of the atmosphere is to run an ensemble of forecasts. In this case, instead of producing a single prediction from an NWP model through a forecast period, an operational centre runs a group or *ensemble* of forecasts through the forecast period, which are all perturbed in some way. The usual way of introducing perturbations into the ensemble members is through the initial conditions, with perturbations which are small enough to be within the limits of observational uncertainty but are still large enough to lead to significant divergence between ensemble members through the forecast period. More recently, operational centres have also introduced perturbations in the formulation of the model physical parametrizations to account for the uncertainty in this aspect of the numerical models. An ensemble needs to include enough members to lead to sufficient spread within the ensemble and this makes ensemble prediction inherently expensive. Only about 10 forecasting centres run ensembles operationally, although this number has been increasing rapidly in recent years.

Ensemble methods have three major advantages over running a single, deterministic forecast:

1. *Quantification of uncertainty.* The main purpose for running an ensemble of forecasts is to try to sample the potential spread in the outcome of a forecast due to the uncertainty in specifying the initial conditions coupled to the non-linearity of the atmosphere. Most ensemble prediction methods are designed to try to maximise the spread of the forecast in order to give forecasters an idea of how unpredictable the atmosphere is during the forecast period. An ensemble of forecasts which all give a very similar prediction could indicate that the atmosphere is in a reasonably predictable state, or it could mean that the perturbations introduced into the individual ensemble members are either not big enough or have not been carefully formulated to fully sample the uncertainty.

2. *Generation of probability forecasts.* With an ensemble of forecasts it is possible to translate the different predictions into probabilities. For instance, if 75% of ensemble members predict rain at a particular location over a 12-hour period whereas 25% predict no rain, a forecast could be issued giving a 75% chance of rain at that location. This translation of ensemble behaviour into a probabilistic forecast is not without its difficulties and dangers, as inherent biases within the numerical model may lead to misleading forecasts. In addition, not all forecast customers welcome a probabilistic forecast.

3. *Improved accuracy.* Whilst it is true that ensemble methods themselves will not make each forecast more accurate, it is also true that, averaged over a reasonable period, the ensemble mean (i.e. the conditions obtained by averaging together the forecasts of all the ensemble members) *will* be more accurate than a set of single forecasts run with the same numerical model over the same period. This is because the ensemble mean will tend to filter out the more unlikely outcomes. It is important to realise that the ensemble mean will not necessarily be the most accurate prediction for each individual forecast period but, averaged over time, the ensemble mean will generally outperform any individual member.

5.2.1   *Ensemble methods*

There are a number of considerations in the design of an ensemble prediction system which are unique to ensemble forecasting – that is they have no counterpart in deterministic forecasting using a single model run. The first and perhaps most obvious consideration is how many members to include in the ensemble. An obvious answer is perhaps 'as many as possible', but this itself isn't easy to quantify, with limits imposed both by available computing power and the methods used to determine how to perturb the individual ensemble members. The second consideration, leading on from this, is exactly how to introduce perturbations into the individual forecast members in a way which will lead to the greatest benefit. We will look at this issue first.

Modern data assimilation methods are designed to give the best possible estimate of the initial state of the atmosphere prior to running an NWP forecast of how that state will evolve. Given that this is the case, how do operational centres then introduce perturbations to those conditions which will mean that the outcomes of the forecasts produced by the perturbed members are equally as likely as that produced by the unperturbed control forecast? It is clear that in order that the perturbed forecasts are equally as likely as the control, the perturbations to the initial conditions must be within the bounds of the estimated error in the analysis, which itself is a factor of the uncertainty in both the observations and the model background field.

The perturbations need to be internally consistent so that a perturbation to the wind field is balanced by perturbations to the pressure and temperature fields. This ensures that perturbations will not lead to high frequency gravity waves as the model dynamics attempt to remove any artificial imbalances in the initial state. It is also important that the spatial structure of the perturbations, in both the horizontal and vertical, is similar to the structures of the typical forecast errors rather than being purely random. Most operational centres also use methods which attempt to introduce perturbations that have structures which map onto the patterns of the fastest growing disturbances in the atmosphere on the day of the forecast. For instance, on a day with a strong mid-latitude jet, the fastest growing disturbances are likely to be baroclinic waves associated with upper tropospheric troughs, ridges and jet-streaks. It makes sense, therefore, to introduce perturbations into the initial conditions which have structures that map onto this type of disturbance. That way the ensemble will give some indication of the uncertainty associated with this type of feature. If perturbations targeted at an upper level trough lead to 75% of the ensemble members developing a deep surface low which did not develop in the control forecast, this is extremely useful information for forecasters who can then factor in the probability of such a development into their forecasts for customers. In the tropics the fastest growing disturbances are tropical depressions, which may or may not develop into full-blown cyclones and whose tracks can be rather unpredictable. Hence, it makes good sense to target some perturbations at any tropical storms in the analysis to get an idea of the uncertainty associated with them.

Breeding methods

The breeding cycle method of obtaining perturbations for an ensemble of forecasts was designed by Toth and Kalnay (1993, 1997) to try to 'breed' meteorologically relevant perturbations to the initial analysis of an NWP model so that they would map onto the fastest growing disturbances present in that analysis. This method is based around short runs of a forecast model into which an initially random perturbation (or 'seed') has been introduced at the initial time. After a set length of time, say six hours or one day, the perturbed forecast is compared to the unperturbed forecast and the difference between the two is re-scaled to match the size of the initial perturbation. This re-scaled difference is then added to the model state and the model is run on again for another period, after which the difference is again calculated and rescaled. This process of comparing the perturbed and unperturbed forecasts and then adding the scaled difference back to the model state is continued for several days, at which point it is found that the differences between the perturbed and unperturbed forecasts start to grow more rapidly. This is because the differences between the perturbed and unperturbed forecasts which do *not* resemble rapidly growing structures have decayed away to negligible size. The remaining differences are *bred vectors* which resemble the fastest growing modes of instability, hence the more rapid divergence from the unperturbed forecast.

   This breeding cycle method is well adapted to an operational forecast model which is running regularly, say once or twice a day. The method does not incur any additional cost apart from running the ensemble itself, and the breeding vectors can be maintained and refined across each subsequent run of the model. NCEP in the USA runs an ensemble using the breeding vector technique to generate the perturbations.

Singular vectors

This approach to generating perturbations for the initial conditions of a forecast ensemble is quite different to the breeding approach described above. In this case, a linearised version of the forecast model and its adjoint (Section 4.3.4) are used to identify the regions of the atmosphere which are most unstable and the structures which would lead to the fastest growing modes of instability. Usually this involves about three forward and backward integrations of the model through a period of 36 or 48 hours. The structures which emerge from this process are called *singular vectors*. Clearly this method involves extra computing expense and a larger time overhead than the breeding vector technique which occurs as the numerical model runs through an operational forecast cycle.

The structures which emerge as singular vectors are often, in mid-latitudes, perturbations on upper level troughs and jet streams at the level of the jet stream maxima. In the tropics they tend to map onto tropical storms or other shortwave features.

ECMWF uses singular vector methods to compute the perturbations for its ensemble forecast. Fifty singular vector structures are selected in each hemisphere, with an additional five singular vectors per active tropical storm (up to a maximum of six storms). These singular vectors are then combined linearly to give 25 perturbations to the initial conditions of the forecast. Each vector is assigned a weighting factor drawn from a random number distribution. These 25 perturbations are then both added to and subtracted from the initial conditions for the control forecast to give a 51-member ensemble (including the control forecast itself). Figure 5.3 shows a typical perturbation to the 700 hPa temperature field which results from this linear combination technique. Although this figure only shows the perturbations at one level of the atmosphere, these patterns also have a vertical structure associated with them. In the same way that increments added by the data assimilation scheme of an NWP model need to be balanced in all the model fields (i.e. the mass and momentum fields must be in agreement with each other), the perturbations to the initial conditions of

21 March 2012 00UTC ECMWF Temperature at 700hPa
Difference between the control and ensemble member 1 (every 0.4K)

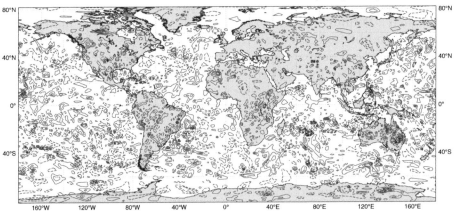

Figure 5.3 Perturbations to the initial conditions of a single ensemble member in the ECMWF ensemble forecast. The field shown is 700 hPa temperature and the contour interval is 0.4K. (Reproduced by permission of ECMWF.)

the ensemble members must also be physically consistent across all model variables.

In addition to perturbations in the initial conditions, ECMWF also introduces perturbations to the model physics of its ensemble members. This is done by multiplying the increments calculated by the physical parametrizations by a random number drawn from a carefully characterised Gaussian distribution with a mean of 1.0 and a standard deviation of 0.2. This helps to increase the spread of the ensemble towards the size of the forecast model errors, which, as discussed above, is a key goal of ensemble prediction.

5.2.2   Operational considerations for ensemble prediction

Because of the expense involved in running an ensemble the number of ensemble members is usually limited to less than would be considered ideal to ensure that the ensemble fully captures the range of potential forecast outcomes at every run of the forecast. Even in the ECMWF ensemble with its 51 members there are times when the actual evolution of the atmosphere falls outside the spread of the ensemble. NCEP runs an ensemble with 21 members (including the unperturbed control forecast). ECMWF runs its forecasts out to 14 days whereas NCEP runs to 16 days. The UK Met Office also runs an ensemble but this is designed to focus on the shorter forecast range and only goes out to 72 hours with 25 members. All operational centres running ensembles do so at lower resolution than the deterministic control forecast. Typically, ensemble members are run at half the resolution of the deterministic control model. Again this is a measure designed to make best use of limited computing resources.

One of the key issues with operational ensemble prediction is how to present the vast amount of forecast information in a way that can be rapidly assimilated by forecasters. It is pretty clear that simply duplicating the output format of a single deterministic forecast for each ensemble member is not the most efficient or intelligent way of presenting the output from an ensemble of forecasts. The three advantages of running an ensemble discussed above are that (i) the ensemble mean is, on average, a more skilful forecast than any single deterministic forecast, (ii) the ensemble spread gives an indication of the predictability of the atmosphere and (iii) the availability of an ensemble of forecasts allows the generation of probability-based forecasts. It makes sense, therefore, that the presentation of ensemble forecast output should be designed to emphasise these three attributes. Figures showing the ensemble mean of particular meteorological fields are therefore commonplace. A wide range of different ways of presenting the ensemble spread usually constitutes the majority of output from an ensemble forecast. Finally, figures which indicate the probability of particular events occurring also constitute a fairly large part of the output from an ensemble. These areas are discussed in more detail in Section 5.3.

5.3   Model output – what can NWP models produce?

The reason for running NWP models is to produce forecast information that can be used by meteorological services to create weather forecasts for their customers. A lot of effort, therefore, is put into the creation of model output formats which can relay the forecast from the NWP model to human forecasters in some sort of optimum way.

The internet and high-speed broadband data transmission technologies have vastly increased the amount of information that can be provided to forecasters at locations away from the centre where the computer model is actually run. As recently as the 1980s it was often rare for forecasters at outstations such as military and civil airports to see very much output from NWP models at all, and often what was transmitted was often too late to have any real use in forecast production. Instead outstation forecasters relied on telephone conferences with senior forecasters at the central forecasting office to provide them with guidance on what the NWP models were predicting. Nowadays pretty much all the information that is seen by the senior forecasting team can be transmitted to outstation forecasters in only slightly longer than it takes for the information to be produced by the models. To some extent this, together with improvement in the skill of NWP models over recent decades, has resulted in a complete change in the attitude of outstation forecasters towards NWP output. With only limited, second-hand and often rather late access to NWP products, outstation forecasters were often reluctant to place too much faith in the forecasts from the models, relying instead on extrapolation-based and empirical forecast methods. Nowadays, NWP is firmly at the centre of all forecast production, both in central forecast offices and on remote outstations.

An NWP model can produce a forecast of pretty much any meteorological variable you could think of (and quite a few that you would not think of at all!). A typical operational forecast model may produce of the order of a hundred or so different forecast fields – usually referred to as *diagnostics*. However, there is often a much smaller set of diagnostic fields that a forecaster or forecasting team will make regular use of. This set will probably vary from location to location, depending on the typical weather systems or the types of forecast being produced. Obvious things that forecasters might want to look at are maps of precipitation intensity, wind speed and temperature through the forecast period, but these fields by themselves are often not enough.

Forecasters want to know the reasons for changes to the weather throughout the forecast period, so want to look into the causes of these changes rather than just simply reporting what the NWP model is predicting. Understanding the reasons behind changes in the weather helps forecasters to understand the degree of uncertainty in the forecast and consider the potential for alternative evolutions of the atmosphere. So even if forecasters are writing forecasts of the weather as it will be experienced at the surface, they often look at predictions of how the mid and upper levels of the troposphere are expected to evolve, as these are the levels which dictate the development of weather systems. Changes to the speed and direction of jet streams, intensifications or weakening of mid-level ridges and troughs all have an effect on the weather that develops at the surface, so forecasters, particularly those in the senior forecasting team within an organisation, will want to have a feel for the development of these types of features before focusing on the detail of the weather conditions at the surface. Other forecasters within the organisation may not have time to do this type of in-depth analysis of the developing situation and will spend more of their time looking at the variables that their customers want to know about. Chapter 6 describes in more detail the processes that senior forecasters go through in assessing the output produced by an NWP model.

For many forecasters, especially those working in mid to high latitude locations, a set of forecast mean sea level pressure (mslp) charts running through the forecast period at 6–24 hour intervals is often the first set of diagnostics that they will consult. To an experienced eye a set of maps of this field will convey a lot of information on the broad state of the atmosphere before going into specific details. The general changeability of the weather over the forecast period will be apparent and pressure maps give a good indication of the predicted strength of the near surface winds. The direction of the flow will give indications of the air mass type, from which qualitative deductions can be made about the forecast temperature and general stability of the atmosphere. In the tropics, surface pressure variations are generally rather small and a set of near-surface streamlines may replace the mslp maps as the diagnostic that conveys the broad-scale picture of the forecast.

Often forecasters will want to inter-compare maps of more than one meteorological variable at the same time. For instance, a map of the upper level geopotential height field (at, say, 300 hPa) overlaying the low level temperature pattern (at, say, 850 hPa) can convey a lot of information on how weather systems will develop. Features at the jet stream level often engage with lower level temperature gradients to enhance or weaken a developing depression. Spotting these types of development in the NWP forecast and having a feel for the uncertainty associated with them allows forecasters to develop a greater understanding of what the model is predicting. Figure 5.4 shows a chart of the 300 hPa geopotential height field and 850 hPa wet-bulb

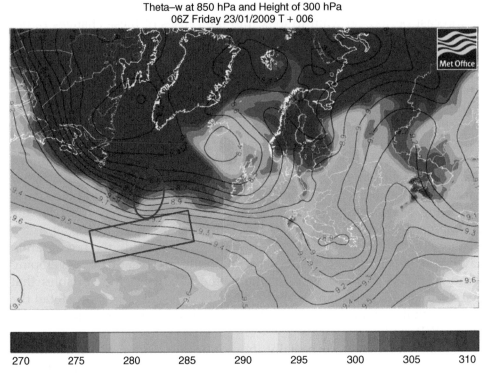

Theta–w at 850 hPa and Height of 300 hPa
06Z Friday 23/01/2009 T + 006

| 270 | 275 | 280 | 285 | 290 | 295 | 300 | 305 | 310 |

Figure 5.4   300 hPa geopotential height (contours) and 850 hPa wet-bulb potential temperature (colours) from the T+6 frame of a UK Met Office global model forecast. The red oval shows the position of the upper trough in the 300 hPa contours and the red rectangle shows the streamer of high 850 hPa wet-bulb potential temperature. (© Crown Copyright 2009, Met Office.)

potential temperature field from the T+6 stage of a forecast from the UK Met Office global forecast model. There is a strong jet stream across the Atlantic, indicated by the closely packed geopotential height contours. A slight trough is discernible in these contours in the centre of the Atlantic, annotated with a red oval. Just to the east of this upper trough is the tip of a streamer of high wet-bulb potential temperature air, annotated with a red rectangle. The presence of an upper trough in close proximity to warm moist air in the boundary layer indicates high potential for the development of a depression.

Forecasters often use satellite imagery to get an overview of the general meteorological development in their area of interest. An experienced forecaster can use a combination of different channels of imagery to locate features, such as shortwave upper troughs, jet streaks and so on, and compare these with the forecasts of the same features from the NWP model over the first few hours of the forecast period. This type of comparison will always be subject to a degree of interpretation of the imagery by the human eye.

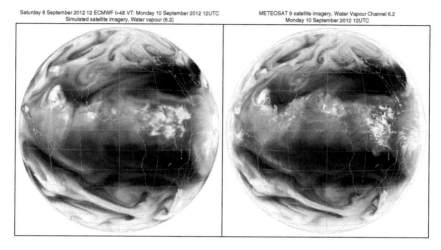

Saturday 8 September 2012 12 ECMWF t+48 VT: Monday 10 September 2012 12UTC
Simulated satellite imagery, Water vapour (6.2)

METEOSAT 9 satellite imagery, Water Vapour Channel 6.2
Monday 10 September 2012 12UTC

Figure 5.5 Simulated (left panel) and real (right panel) water vapour satellite images. The simulated image was generated by the ECMWF global model and is a T+48 forecast. (Reproduced by permission of ECMWF.)

To overcome this problem, it is possible for an NWP model to produce forecast satellite images that can be compared more directly with the actual imagery to spot similarities and differences. Using the cloud predicted by the NWP model together with the radiative transfer scheme of the model, it is possible to simulate directly what a satellite in a particular orbit, viewing the Earth at a particular wavelength, would actually be seeing. Figure 5.5 shows a T+48 forecast water vapour (WV) channel satellite image from the ECMWF NWP model, together with the actual WV image from Meteosat at the verifying time of the forecast. We can see that the overall large scale patterns predicted by the model all look pretty good. However, there are a number of smaller differences between the two pictures. However, the model has not done a perfect job at predicting the convective activity in the Intertropical Convergence Zone (ITCZ), particularly over West Africa.

Having assessed the broad-scale trend of the forecast, using a combination of mid-to-upper tropospheric and surface diagnostics, forecasters will then generally move on to look at more specific aspects of the forecast in order to start producing information for their customers. Here the possible diagnostics are determined largely by customer requirements. Bear in mind that an NWP model can produce a forecast of pretty well any meteorological variable you could think of. As well as the obvious ones such as precipitation, temperature and wind speeds, more application-specific fields such as soil moisture content, cloud ice content, water held on the vegetation canopy and surface evaporation rates are all fields that have a potential customer, and hence can be produced by an NWP model.

5.3.1   Ensemble diagnostics

As stated above, a major issue with ensemble forecasts is that they generate vastly more output than a single deterministic forecast. This presents the problem of how to display all the extra output in a way that can be assimilated quickly by a team of forecasters and which emphasises the attributes that make running ensembles a useful exercise in the first place.

The first reason for running an ensemble of forecasts is to generate an assessment of the predictability (or uncertainty) of the meteorological situation. Hence, a large number of ensemble diagnostics are designed to convey information to help forecasters quantify this.

A very simple first diagnostic aimed at doing this is the so-called 'postage stamp plot', showing the output of all the ensemble members in one particular meteorological field (often mean sea level pressure) for a given lead time in the forecast over the area of interest. Figure 5.6 shows one such plot from the ECMWF ensemble for a particular date in March 2004, 132 hours (5.5 days) into the forecast period. All 51 members (plus the high resolution control forecast) show an area of low pressure over Eastern Europe with a ridge of

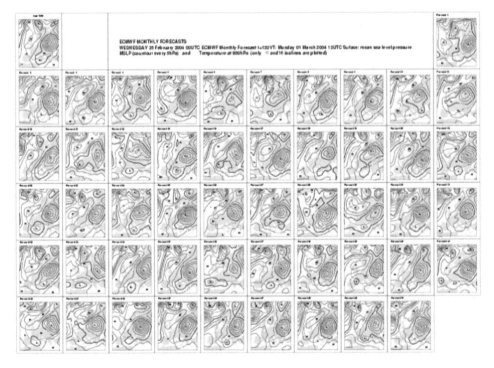

Figure 5.6   A 'postage stamp' map of mean sea level pressure over Europe on 1 March 2004 from the ECMWF ensemble. Each frame shows the T+132 forecast from one of the 51 ensemble members. (Reproduced by permission of ECMWF.)

high pressure in the west, near the United Kingdom and extending across Scandinavia. There are large variations between ensemble members as to the central pressures and exact locations of these two main systems but the overall pattern is similar across the whole ensemble, implying some degree of predictability in the forecast. A forecaster will not be looking at the specific detail of each forecast but instead will be looking for broad similarities and differences across the whole ensemble.

The postage stamp plot allows a forecaster to make a subjective assessment of the similarities and differences between the forecasts from the different ensemble members. This type of assessment can be made more objective through a technique known as 'clustering'. Clustering is a ubiquitous concept in statistical data analysis whereby individual data points within a set are grouped together in some way based on an objective definition of similarity. The whole data set can then be represented by the means of the various groups or 'clusters' rather than the individual samples, thus reducing the overall amount of information that needs to be assessed by the user. In meteorology, clustering is usually performed by applying some kind of pattern recognition algorithm to a particular forecast variable (say, 500 hPa geopotential height) over a particular region of interest (say, Europe or the USA) at some particular lead time (say, days 3–5 in a 10-day forecast). The clustering algorithm will group together forecasts which are deemed to be similar in a pre-defined way and will then compute means of all the ensemble members in each cluster. Typically the clustering algorithm will be tuned to produce between one and about five clusters, with the number of clusters being some sort of indicator of the uncertainty inherent in the forecast. A single cluster indicates a reasonably high degree of predictability whereas an increasing number of clusters indicates increasing uncertainty in the forecast. The number of ensemble members within each cluster also helps forecasters make some assessment of the predictability. Even if there are several clusters, if the vast majority of ensemble members sit in the main cluster, with the others perhaps only containing two or three members each, then a forecaster can still express most confidence in a forecast outcome based on the most populated cluster. A forecast which produces two or three clusters all containing a similar number of ensemble members indicates more uncertainty.

A set of cluster means is shown in Figure 5.7. The field shown is mean sea level pressure but the clustering algorithm was actually performed on 500 hPa geopotential height. The main difference between the clusters is the position and shape of the depression in the eastern Atlantic. Cluster 0, with approximately 51% of the ensemble members, shows a rather slack pressure gradient over much of the United Kingdom and Ireland, whereas cluster 1, with 26% of the ensemble members, shows a stronger southerly flow over Ireland and the western United Kingdom. Cluster 2, with 11% of the

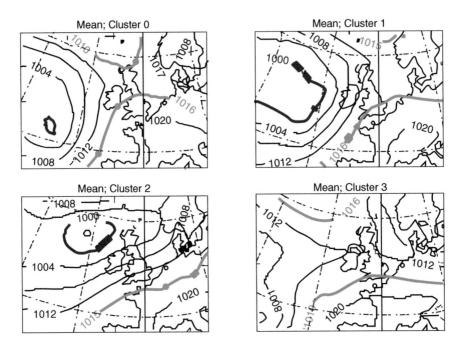

Figure 5.7 A set of cluster means from an ECMWF forecast over Western Europe and the eastern Atlantic. The field shown is mean sea level pressure but the clustering algorithm was based on 500 hPa geopotential height. Cluster 0 contained 26 of the 51 ensemble members, cluster 1 contained 11 members, cluster 2 and cluster 3 both contained seven members. (Reproduced by permission of ECMWF.)

ensemble members, has the depression further north and east, resulting in a significant south-westerly flow over most of the United Kingdom and Ireland whereas cluster 3, also with 11% of the ensemble, has the low much further west and a very slack pressure 'col' over the United Kingdom. Clusters 0 and 3 between them contain over 60% of the ensemble members and both indicate rather slack pressure gradient across the United Kingdom, which could be interpreted as the most likely outcome. However, with both clusters 1 and 2 containing a significant number of ensemble members, the possibility of a stronger flow across parts of the United Kingdom and Ireland cannot be ruled out in this forecast. For specific details regarding the ECMWF clustering methodology and output, see Ferranti and Corti (2011).

Verification statistics show that, averaged over a long enough period, the ensemble mean is a more skilful forecast than any of the individual ensemble members, for the reasons discussed in Section 5.2.1. However, it is rather unusual for forecasters to look purely at maps of the mean of the whole ensemble for any given individual forecast period. This is because the ensemble mean by itself contains no information about uncertainty or any potential alternative scenarios. Additionally, the ensemble mean will tend towards a rather bland featureless set of diagnostics as all the variability and sharp gradients will tend to be smoothed out. If the ensemble mean of a particular field is used as a diagnostic it is usually plotted with shading that indicates the spread of the ensemble in some way (often the standard deviation across the ensemble) in order to convey some information on the uncertainty within the ensemble. Looking at *cluster* means gives forecasters much more information than simply looking at the mean of the whole ensemble. Of course, if all the ensemble members produce a very similar forecast (as defined by the pattern matching algorithm used in the clustering process) then they may well all fall into the same cluster, the mean of which *will* be the ensemble mean.

Any kind of averaging, be it over the whole ensemble or smaller groups, will tend to smooth out information about extreme values. Whilst forecasters want to know about the most likely outcome of a particular forecast, they would also like to know something about the possibility of the occurrence of unusual or extreme conditions as these are likely to have the biggest impact should they actually occur.

The predictability in any weather forecast will almost certainly be geographically dependent, so it is useful to have a set of diagnostics which convey information on this variability. The most widely used method of doing this is the use of the so-called 'spaghetti plot'. For a field such as mid-tropospheric geopotential height or atmospheric thickness a couple of key contour values are chosen and those contours are plotted for each ensemble member on a single map covering the region of interest. The resulting plot can be a rather messy set of lines (hence the name) but there will probably be parts of the map where the contours from each ensemble member are rather close together, and other parts where they are spread over a wide area. Figure 5.8 shows an example of such a plot from the ECMWF ensemble for a date in February 2012, at a lead time of six days. The field plotted is 500 hPa geopotential height with the orange contours showing the 5700 m height and the blue contours showing the 5200 m height. The green and dark blue contours are from the control member (i.e. with unperturbed initial conditions); the purple and red contours are the values from the verifying analysis which have been added after the fact for verification purposes. Over North Africa and central Asia the 5700 m contours are all closely bunched, indicating good agreement between the ensemble members in this region. However, over the southern USA and Northeast Canada/Greenland the 5200 m contours are a real mess of lines, indicating a degree of uncertainty in the position and depth of the trough in this region. Some ensemble members, including the control forecast, are indicating a cut-off feature centered on Baffin Bay whereas other members have a fairly zonal pattern in this region. Clearly, the forecaster cannot look at the detail of every single member on a map like this but the mass of messy lines gives a good visual indication of uncertainty in the forecast in the North American sector.

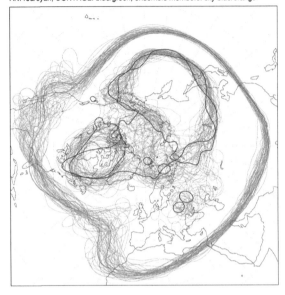

5 Feb 2012 00UTC ECMWF ensemble forecast t+168h VT: 12 Feb 2012 00UTC
500 hPa Geopotential (520/570 isolines plotted)
AN: red/cyan, CONTROL: blue/green, ensemble members: sky blue/orange

Figure 5.8 A 'spaghetti plot' of 500 hPa geopotential height at T+168 hours from the ECMWF ensemble on a day in February 2012. Orange contours are 5700 m and light blue contours are 5200 m. The green and dark blue contours are from the control member in the ensemble and the cyan and red contours are from the verifying analysis. (Reproduced by permission of ECMWF.)

The diagnostics described so far are all based on spatial maps. Ensemble products also bring a useful added dimension to location specific diagnostics. 'Plume' plots for individual locations such as cities are widely used to convey the uncertainty in particular fields at specific locations through the period of a forecast. In this type of plot the forecast values of a particular variable at a specific location (or the nearest available model grid point) are plotted against time for every ensemble member. Similar to spaghetti plots, a tight grouping of all the lines implies a reasonable degree of predictability in the forecast whereas a wide spread indicates a high level of uncertainty. In general, the spread will tend to increase through the forecast period. However, sometimes this transition is not gradual but happens rapidly over a relatively short period, implying a rather rapid change from predictable to unpredictable conditions at that point in the forecast. Figure 5.9 shows plume plots of 2 m temperature in London from the ECMWF ensemble for forecasts initiated on 26 June in 1994 and 1995. The red lines show the forecast of each individual ensemble member and the thick black line shows the control forecast. The dashed blue line, added after the forecast, shows the verifying analysis values. The forecast for 1994 implies a greater degree of predictability than that for 1995. Note that in 1994 the spread in the ensemble increases

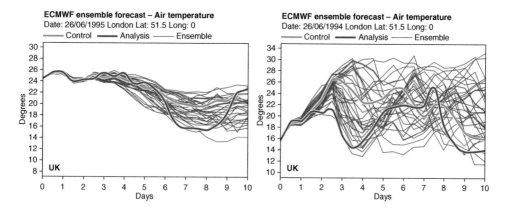

Figure 5.9 Plume plots of 2 m temperature in London from two forecasts from the ECMWF model. Red lines show the forecast for each individual member and the thick black line in each case shows the forecast from the control member. The dashed blue lines show the verifying analysis. The labels on the y-axis are in degrees celsius. (Reproduced by permission of ECMWF.)

rapidly after about 2.5 days into the forecast whereas in 1995 the spread increases gradually through the whole forecast period. In both cases the actual temperature is within the spread of the plume but in 1994 particularly, the majority of ensemble members are predicting temperatures which are rather too high, implying a degree of warm bias in the model.

The use of location-specific forecast diagnostics can be taken further with *meteograms* (or *epsgrams*). These also use the format of plotting a given variable against time at specific locations, but instead of showing each ensemble member as a line, the spread of the ensemble is shown using 'box-and-whisker' icons every 6 or 12 hours through the forecast period. This type of plot, as well as conveying information on the uncertainty in the forecast, allows the generation of probability forecasts. The box-and-whisker icons show the percentage of the ensemble members predicting values in a certain range, or above or below a certain value. Assuming an unbiased forecast, these percentages can be translated into probabilities – that is if 90% of the ensemble members predict a temperature of greater than 20°C then this can be interpreted as a 90% probability of temperatures exceeding 20°C. Of course, this translation of the ensemble distribution into a probability forecast needs to be used with great care to allow for known model biases.

Figure 5.10 shows a meteogram from the ECMWF ensemble prediction system for a location in Finland. The four panels show cloud cover, precipitation, 10 m wind and 2 m temperature at six-hour intervals through the period of a 10-day forecast. The ends of the whiskers in each icon show the extreme values predicted by the ensemble. The ends of the thinner blue boxes show the 10th and 90th percentiles (i.e. 80% of the ensemble members lie between these two values) and the ends of the thicker blue boxes show the 25th and

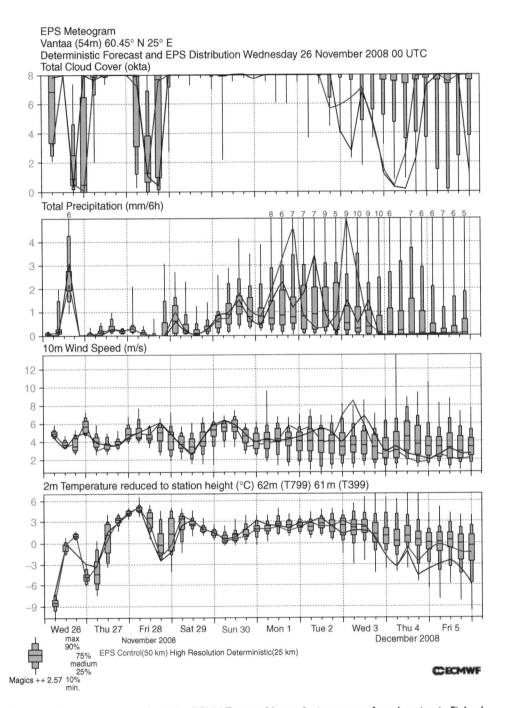

Figure 5.10   A meteogram from the ECMWF ensemble prediction system for a location in Finland. The panels show, from the top, cloud cover, precipitation, 10 m wind speed and 2 m temperature. The box-and-whisker icons are described in the text. (Reproduced by permission of ECMWF.)

75th percentiles (i.e. 50% of the ensemble members lie between these values). The horizontal line near the middle of the thick blue box shows the ensemble median value (note that this is *not* the ensemble mean). The continuous red and blue lines in these figures show the predictions from the ensemble control member and the high resolution deterministic model respectively. Although the ensemble spread generally increases through the forecast period, there are some interesting features in the individual variables. In the cloud forecast there is a large degree of spread in the ensemble at the start and towards the end of the forecast, but from day 4 to day 7 all or almost all of the ensemble members are predicting total cloud cover. This confidence in the ensemble is also reflected in the prediction of 2 m temperature which also shows a very small spread through day 4 to day 7. In precipitation the spread tends to increase through the forecast period, but the spread is also quite large during the predicted rain event on day 1. Note, however, that every single ensemble member is predicting at least 1 mm of rain during this event, indicating a high degree of confidence that there will be a rain event during this period.

The translation of the distribution of an ensemble of forecasts into a probability forecast is a major reason for running ensembles, as discussed in Section 5.2.1. Hence, there are many different diagnostics which can be used to convey probabilities of particular events. Figure 5.11 shows one way of doing

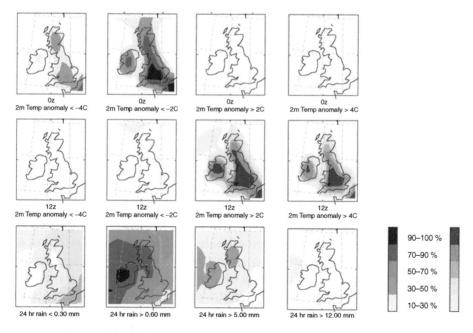

Figure 5.11   Probability forecasts from the ECMWF ensemble for 00:00 UTC temperature anomaly (top row), 12:00 UTC temperature anomaly (middle row) and 24-hour precipitation (bottom row). (Reproduced by permission of ECMWF.)

this. For one day during the forecast period, maps of the United Kingdom are plotted for 00:00 UTC temperature (top row), 12:00 UTC temperature (middle row) and 24-hour precipitation accumulation (bottom row). For each variable there are four panels plotted, indicating the probability of that variable either being below or exceeding a given value. For instance, the top left map shows the probability that the 00:00 UTC temperature will be more than 4°C below normal.

To define 'normal' for these probability forecasts, many years of model hindcasts have to be run to generate a model climatology for each variable. This is because, unless the NWP model is absolutely perfect, it will have a different climatology and, hence, a different value of 'normal' than the real atmosphere. Not only that but the model climatology will be forecast lead-time dependent. Over the period of a forecast the state of the atmosphere in an NWP model will tend to 'drift' towards a more unrealistic state. At the start of each forecast the data assimilation pulls the model back towards a correct state. So the model climatology for the first 24 hours of the forecast will be closer to the real atmosphere than for the 24–48 hour period, which in turn will have a more realistic climatology than the 48–72 hour period and so on. Only by running many years of hindcasts can the forecast lead-time dependent climatologies of the model be correctly quantified so that they can then be used when producing forecasts of a particular variable relative to 'normal'. This eliminates the model bias, as the forecasts are showing the probability of that variable being different from the model climatology as opposed to the observed climatology. This obviously vastly increases the expense of generating these types of diagnostic, and every time the model is changed the hindcasts need to be re-run in order to generate the new model climatologies.

5.3.2   Model output post-processing

NWP models can produce a vast amount of diagnostic information and if an ensemble of forecasts is run even more information can be produced. Even so, a numerical model (or single member of an ensemble) can never do more than produce one forecast of a particular variable *at each grid point of the model*. If the model covers a large domain, and hence has a grid box size of perhaps 25–50 km square, then a single value of, for instance, overnight minimum temperature may mask a lot of local variability. This is particularly true if that grid box contains a land region with appreciable local topography and maybe also contains contrasts between rural and urban areas. Even in a high resolution regional model with grid boxes perhaps 5 km square, a single value of some variables may still mask quite a lot of local variation. The obvious answer to this problem is to increase the resolution of the model so that the local variations in terrain and land surface type which lead to the variations are themselves resolved. This is a somewhat impractical solution as most NWP centres are already running their NWP systems at the highest resolution that is allowed by the constraints of the available computing power. A more pragmatic solution is to put the output of the NWP model through some sort of further processing in order to incorporate some local variability. This approach is usually referred to as 'post-processing'.

Model output statistics (MOS)

NWP models are generally quite good at predicting the large scale weather conditions such as the pressure distribution, upper level geopotential height fields and the temperature of the atmosphere above the boundary layer. However, they are much less skilful at predicting small scale variations in weather, particularly near the surface. To many users of weather forecasts, it is these small scale variations that are most important to them and so forecasting agencies have developed ways of producing forecasts of these local variations based on the large scale patterns that the models are known to be reasonably good at forecasting. These types of methods are usually referred to as Model Output Statistics, or MOS forecasts.

MOS methods are based on having reasonably long records of both forecast variables from the NWP model (the 'predictors') and observations of the variables that need to be forecast at the specific locations where forecasts are required (the 'predictands'). Statistical relationships can be derived between predictors and predictands. Usually, several different predictor variables are combined using multiple regression techniques to make the MOS forecasts. For instance, to predict overnight minimum temperature at a given location, the 850 hPa temperature, 10 m wind speed and cloud cover forecasts from the NWP model might be combined into a statistical model that gives the best fit to the observed overnight minimum temperature at a given location. Hundreds of these models can be built and run using automated software and the models can be refined every time a new forecast and a new set of observations are added to the database. Although these statistical methods are fairly complex, compared to the complexity of a full NWP model they are extremely cheap to run and generally add skill to the forecast. There will, of course, be circumstances in which the derived statistical relations between predictors and predictands break down but over a reasonable period MOS forecasts are generally an improvement on raw NWP output. MOS can also be used to generate probability forecasts. These are based on the past occurrences of a particular weather condition given the same large scale situation. This is a very different method to the use of an ensemble to generate a probability forecast, as it relies on a long data set of past forecasts. Again, there is the potential problem of unusual circumstances, which have not been seen many times before in the past record, leading to incorrect forecasts but on the whole MOS methods for generating probability forecasts do have skill and they have been the basis for probability forecasting, particularly in the United States, for many years.

Site-specific forecast models (SSFMs)

MOS uses statistical methods to put localised detail into raw NWP forecasts. A rather more physical approach is the use of site-specific forecast models (SSFMs). These models explicitly take into account the local variations in terrain and land surface type when producing forecasts. Effectively, a SSFM is a numerical model of the column of the atmosphere containing the particular location of interest, with an increased vertical resolution in the boundary layer and very high resolution information (of the order of 25 m) on the nature of the land surface. The model is 'driven' by the weather taken from the corresponding column of the full NWP model; this is then modified, largely by the interaction with the underlying surface. The enhanced vertical resolution together with the additional information on the nature of the underlying surface allow more accurate prediction of the boundary layer, particularly in situations where the surface layers become 'de-coupled' from the free atmosphere, such as very light wind conditions and overnight under the influence of a temperature inversion. SSFMs have proved to produce more accurate forecasts of fog formation and overnight temperature variations than the full NWP models which provide the driving information. In contrast to MOS, they are not susceptible to forecasts going wrong due to the occurrence of a set of weather conditions which have not occurred very often in the past record. However, they are more expensive to run and as NWP model grid boxes decrease in size to be of the order of 1 km, the benefits of running SSFMs is reduced as the full NWP model becomes better able to capture the small scale variability.

5.4   Using NWP output to drive other forecast models

Site-specific forecast models are just one example of models which take output from an NWP model and process it in order to produce a new forecast. There are many other examples of this type of modelling being used to produce specialised forecasts for particular customer groups. Some of these models may be run by the forecasting agencies themselves, whereas others may be run by other agencies or companies who have an agreement with the forecasting centre to use their model output. Power generation companies, for instance, run models which predict demand for electricity. Much of this demand is determined by human factors but temperature can have a big impact on demand for electricity and gas and so temperatures are a vital input into such models. Health services might consider running models which predict hospital admissions rates during extreme weather events. For instance, high temperatures combined with stable conditions in cities can lead to the trapping of pollution at street level and a consequent rise in hospital admissions due to breathing difficulties and heart problems. A model which can predict this type of event several days in advance can mean that staff can be put on stand-by and extra capacity can be opened up in readiness to deal with increased admissions.

5.4.1 Wave and swell models

Ocean wave and swell heights are of paramount importance to shipping of all sizes and also to operators of offshore facilities such as oil and gas platforms and wind farms. Many large meteorological agencies and smaller specialist firms employ forecasters in commercial ship routing duties and for these forecasters it is wave and swell heights that are more important than atmospheric conditions. For fixed installations, such as drilling rigs, it is often the wave *period* that is more important than its size as all rigs and drilling equipment have a resonant frequency at which vibrations to the structure can amplify rapidly and cause considerable damage. Damage to offshore platforms and large commercial vessels can be extremely expensive, so shipping and drilling companies are prepared to pay considerable sums of money for forecasts which allow them to route their vessels away from potentially damaging ocean conditions or to pull up their drilling equipment prior to the onset of damaging waves.

Specialist models of ocean surface conditions have been developed and are run at a number of meteorological agencies. The main inputs to these models are surface wind and surface pressure predictions from an atmospheric model. Similar models have also been developed for coastal flood prediction, which also need detailed tidal information as it is the combination of storm surge conditions coupled to high (spring) tides that are most conducive to coastal flood risk.

5.4.2 Atmospheric dispersion models

Many meteorological agencies have a responsibility for predicting the dispersion of atmospheric pollutants, both natural and man-made.

Chemical releases, nuclear accidents and volcanic eruptions all inject material into the atmosphere which can have a major impact on human activities and health, so need to be tracked and forecast accurately. To do this type of forecasting, most centres use *Langrangian trajectory models*, which are based on injecting into the atmosphere and then tracking, individual particles, using the wind predictions from an NWP model to move the particles around in the atmosphere, together with information from the model physics schemes to help predict the rate at which particles will be removed from the atmosphere, for example being washed out by precipitation.

One major problem with this type of model can be knowledge about the source of the pollution. In the case of a nuclear or chemical release, the position of the source can be very accurately pinpointed, both in the horizontal and the vertical. In the case of a volcanic eruption, however, the *vertical* distribution of the material injected into the atmosphere may be very imperfectly known, leading to inaccuracies in the forecast of its dispersal through the atmosphere.

Lagrangian trajectory models can also be run backwards in order to trace the source of pollution. In circumstances where pollution sensors are detecting abnormally high levels of a particular species of pollution, particles can be injected into the model at the location of the sensors and the model can then be run backwards in time to try to pinpoint potential sources.

More sophisticated transport models may also include some level of atmospheric chemistry. This allows the prediction of chemical reactions which may occur within the atmosphere transforming a particular pollutant into a different species. This may be particularly important for predicting the occurrence of high levels of pollutants which may be harmful to human health. For instance, high levels of ozone are dangerous to health and may be brought about by the reaction of various oxides of nitrogen with carbon monoxide or other organic compounds.

5.4.3   Hydrological models

Flooding caused by heavy rain, either in the form of flash flooding or brought about by a long period of above average rain, can cause significant damage to infrastructure, buildings and crops and can also endanger human life. Once the water is on the ground, sophisticated hydrological models are needed to calculate whether heavy rainfall will be translated into flooding. These types of model often run on a river catchment basis and include extensive information on catchment characteristics. Local topography, vegetation, soil type and underlying geology all contribute to the run-off characteristics of a river catchment. For instance, the UK Met Office works in partnership with the UK Environment Agency to maintain a national Flood Forecasting Centre

(FFC) for England and Wales. Meteorological and hydrological forecasters work together in this centre to produce forecasts and issue warnings of river and coastal flood risk, and rainfall forecasts from the Met Office UK regional NWP model are used to drive hydrological models maintained and run by the Environment Agency.

5.4.4   Road surface models

In countries which experience cold weather during the winter season, a major customer group for weather forecasts is often the agencies responsible for maintaining safe road travel. Ice and snow on roads is a major hazard, both to the economy and to human life, so some countries make considerable investment in services to keep roads open and safe to travel during cold winter weather. Due to the nature of road surfaces, forecasts of the temperature from even a very high resolution NWP model are often not sufficient by themselves to alert authorities to the potential of frozen road surfaces. The thermal properties of tarmac differ quite considerably from the grid box average surface conditions and even the amount of traffic on a road can considerably change the temperature of the surface through the effects of the friction of tyres and radiation of heat from the hot engines of vehicles. Local topography can also have an effect on roads within a single grid box, with a road in a known frost hollow being more prone to freezing than a location perhaps only a few hundred metres away.

Sophisticated road surface energy balance models can be used to translate temperature forecasts from NWP models into road surface temperature predictions by taking into account the thermal properties of tarmac, the local topographical conditions and the traffic loading on the road in question. Investment in this type of model is worthwhile, as the cost of ice prevention measures, such as applying salt to road surfaces, is very high and the potential costs to the economy and potentially to human life of *not* applying ice prevention measures can be even higher. It is also the case that the cost of preventing ice forming on road surfaces is much lower than the cost of melting ice that has already formed, so responding to a good quality forecast of ice risk is cheaper than simply responding to icy conditions as they start to occur.

In some countries ice prediction models are linked to road surface sensors. These transmit data back to the modelling centres so that forecasts can be constantly adjusted in light of actual observations. Information on the state of the road surface (e.g. wet or dry) can also be sensed and this is crucial information as no matter how cold a road surface is, ice will not form on it if it is dry, unless there is a hoar frost.

Summary

- A number of different considerations determine the nature of the NWP models run by a major forecasting centre. These include the requirements of forecast customers and the availability of computing power. Most major centres run a suite of models to satisfy these considerations.
- When designing NWP models, choices have to be made as to model domain, resolution, forecast range, the amount of detail included in the model physics and the type of output that the model will produce.
- NWP models with regional domains need information from a global domain model (or at least from a regional model with a wider domain) to provide information at the regional model boundaries.
- Ensemble forecasting methods allow the quantification of uncertainty in the forecast and the possibility of producing probabilistic forecasts. Averaged over a long enough period, the mean of an ensemble of forecasts will also be more accurate than a single deterministic forecast using the same NWP model.
- Output from NWP models often undergoes some form of 'post-processing' in order to add more detail or to improve the accuracy of site-specific forecasts.
- Output from NWP model forecasts is used operationally to drive a wide variety of other different types of forecast model.

6

The Role of the Human Forecaster

The preceding chapters have emphasised the importance of state-of-the art technology in modern weather forecasting. For example, sophisticated remote sensing instruments, enhanced bandwidth communications networks and high performance computing facilities all underpin today's numerical weather prediction systems. This may give the impression that human forecasters play a fairly marginal role in forecast production, with forecasts being produced entirely by automated systems. It remains very much the case, however, that human forecasters still play a key role in generating weather forecasts and in then communicating these forecasts to their customers, whether these customers are governments, commercial organisations, the military or the general public. To some extent the advances in automation, high performance computing and communications networks mean that human forecasters actually have more decisions to make than in the pre-NWP era. We will now see how this 'forecaster–machine mix' is harnessed to best effect.

Certainly the role of the human forecaster has changed considerably since the advent of NWP and, particularly, since the availability of high bandwidth communications networks. In the pre-NWP era, weather forecasts were produced by forecasters through careful analysis of the current state of the atmosphere and the application of methods and techniques designed to predict its development, for example using the sort of empirical techniques described in Section 2.3. Meanwhile, in the early days of routine NWP, and even well into the 1980s, the numerical forecasts were limited in forecast range, often of fairly coarse resolution and sometimes subject to fairly significant systematic errors. But very often the main constraint upon the use of NWP forecasts during this period was simply the availability of forecasts to the personnel whose job it was to issue forecasts to customers, particularly if those forecasters were working at a location remote from the central

---

Operational Weather Forecasting, First Edition. Peter Inness and Steve Dorling.
© 2013 John Wiley & Sons, Ltd. Published 2013 by John Wiley & Sons, Ltd.

forecasting office. These forecasters often only had access to a very few products from the NWP system, and these often arrived too late to be of any use when issuing forecasts to customers. Forecast maps were transmitted down low bandwidth telephone lines and often printed by poor quality and very slow facsimile machines. This meant that many decisions had to be made without the aid of NWP products or using NWP guidance that was already rather out-of-date.

Whilst today the raw forecast *is* largely produced by automated NWP systems, human forecasters still have many decisions to make and have a large input into the final form of the forecast as it is communicated in a more digestible form to customers. One big change since the 1980s is simply the volume of NWP output available to forecasters, not just from their own in-house NWP system but also from various other numerical models from forecasting centres around the globe. Internet technology has made it possible for output from all the main NWP centres to be available to forecasters anywhere in pretty much real time. This has had a profound effect on forecasting techniques, particularly in the time range beyond 48 hours. In the past forecasters may have had guidance from a single deterministic NWP model on which to base their forecasts for the medium range. Nowadays, forecasts from multiple global models are available (known as a 'poor man's ensemble'), as well as ensemble forecasts from a range of different centres. The requirement for efficient processing of this increased volume of information has presented new challenges for medium range forecasting and it has certainly increased the complexity of decisions that need to be made by human forecasters working on forecast production at this time range.

In this chapter the role of the human forecaster in the forecast production process is examined. This examination follows two main strands. Firstly, the job of the senior forecast team of a major forecasting centre will be described as it assesses the output from its own NWP systems and also the data available from other forecasting centres. The assessments and decisions made by this senior forecasting team will form the basis of guidance issued to other forecasters, both within the organisation but also possibly working in other organisations or companies. This second group of forecasters has the task of producing and communicating forecasts to customers and it is the role of this group that forms the second strand of this chapter.

6.1   The role of the senior forecasting team

In any large NWP forecasting centre, there will be a team of senior forecasters, working a shift roster, whose job it is to assess the output from NWP models (both those run in-house and from external sources) and to issue guidance to other forecasters who are directly involved in forecast production for customers. This guidance may include an overall summary of the forecast narrative, some indication of the consensus or disagreement between models from different centres and some guidance as to the uncertainty in the forecast.

6.1.1   Short-range assessment of NWP forecasts

Because of the continuous and cyclic nature of weather forecast production (Section 5.1.3), it is rather difficult to define where the job of a senior forecaster actually starts. Perhaps the easiest place to start is the point at which a run of one of the models in the NWP system starts to produce forecast information. For a forecast run with any given initial time, the length of time which elapses between the T=0 time of the forecast and the model actually starting to produce forecast data will be of the order of a few hours – probably between two and six depending on the nature of the model. To give an example, a forecast run with an initial time of 00:00 UTC will start to produce forecast data at perhaps 03:00 UTC. This means that, by the time the NWP model produces a forecast for the first few hours of its forecast period, those few hours have *already happened in reality*. This allows the forecaster whose job it is to assess the quality of each forecast a chance to make a real-time comparison between the forecast and the actual weather over that early period of the forecast. Given the current advanced state of NWP the differences between the forecast and reality in the first 3–6 hours of the forecast period are generally rather small. However, there may be occasions, particularly in synoptic situations where the flow is rapidly evolving, when there may be significant departures from the true evolution of the atmosphere over these short timescales.

A key element to making a comparison between the forecast and reality over the early period of the forecast is to work quickly, with the help of state-of-the-art visualisation tools (e.g. the Hungarian Meteorological Service, OMSZ, HAWK system). Once the forecast computation has completed there will be forecasters who want to use the output in order to produce new forecasts for their customers using the most up-to-date information. This means that the comparison of the forecast to reality is unlikely to be a grid point by grid point comparison of numerical values such as temperature or rainfall rate, but rather a broader comparison of the evolving synoptic situation. The appropriate observational tools for this type of comparison are ones which convey a frequently updating visual picture of the synoptic situation, with satellite and radar imagery being the obvious candidates for this type of work. Using these tools the position of cloud systems and rain-bearing features in the forecast can be rapidly compared with those from the forecast over a fairly broad area. Satellite imagery can also be used to assess the development of key weather features. With imagery from geostationary satellites being available every 15 or 30 minutes, particular cloud features such as fronts or convective systems can be tracked through a sequence of pictures. If the features are getting bigger with time and if, in particular, the cloud tops appear to be cooling with time (as shown by clouds becoming brighter in an infrared image) then it can be inferred that there is significant ascent of air going on within the system which is leading to the clouds deepening. In the case of a depression this implies development and falling surface pressure. In the case of a convective system this would also imply that the system is becoming more intense.

Cloud features can also be used to imply the positions of important features in the upper atmosphere, such as jet streams and ridge and trough axes. These can then be compared with the upper level geopotential height fields from the NWP model to give an overview of how well these features have been represented in the model in the early stages of the forecast. Clearly, this type of work requires a high degree of skill and training in analysing features in satellite imagery. Jet steam positions can be inferred from the position of bands of high cold cloud, often with a sharp edge on the poleward side. The position of upper level trough axes is often indicated by a contrast between deeper convective clouds on the downstream side of the axis with more suppressed convective or stratiform clouds on the upstream side. A quick comparison of the forecast and observed positions of such features will give an initial impression of how the model has handled the first few hours of the forecast period. Large discrepancies are not generally expected but subtle differences in, say, the curvature of an upper trough or the exact position of a jet stream axis may result in differences in the evolution of the synoptic weather features through the forecast period.

On a more local scale, forecasters will be using rainfall radar imagery to compare the forecast distribution of precipitation with the observed pattern. The forecaster will be looking for differences in timing and intensity of precipitation features which may have implications for the forthcoming period of the forecast. A band of rain which is moving slower or faster than in the NWP forecast will need to be adjusted for in forecasts issued to customers, as will the intensity of rain which has been over- or under-predicted by the model. In these situations, as well as using the observed rainfall distribution, a forecaster may also draw on their knowledge of systematic errors in the model. Typical systematic errors in precipitation in a wide range of numerical models include:

- Insufficient orographic enhancement of precipitation rates. Because the orography in NWP models is usually smoother and lower than in reality, orographic enhancement is too weak.

- Lack of inland penetration of maritime showers in an onshore flow during winter. Most NWP models do not advect individual shower features from time step to time step but predict new showers at each time step based on the stability at each grid point. Over land the surface may be too cold to trigger convective clouds but in reality showers which formed over the sea may penetrate some distance inland before dying away.
- Poor representation of rain turning to snow in frontal precipitation. The transition from rain to snow depends very strongly on the temperature structure of the lowest kilometre or so of the atmosphere and the microphysical processes such as the evaporation of rain that occur in this layer. Failure to resolve the fine scale details of the temperature in this part of the atmosphere can lead to errors in the prediction of frozen precipitation.
- Mistiming of the diurnal cycle of convection. Many NWP systems produce convective precipitation too early in the day which then dies away too quickly in the evening. This is because the complex processes of convective organisation are rather crudely represented in convective parametrizations (or not represented at all) so that the diurnal cycle of convection is often too closely tied to the diurnal cycle of solar heating.

As well as these issues, there may be specific problems in particular synoptic situations. All NWP centres will maintain a database of the known systematic errors of their NWP models. Forecasters will be looking out for these issues when assessing the quality of the forecast through its early stages.

Where a human forecaster can really make a big contribution to the quality of the forecast is in the fairly rare cases when the NWP forecast is starting to diverge significantly from reality over the first few hours of a forecast. On a synoptic scale, careful analysis of satellite imagery can indicate the presence of a rapidly developing frontal wave that may not have been well handled in the NWP forecast. On more local scales, rainfall radar images may indicate the development of intense convective precipitation which has been under-predicted or mis-located in the numerical model. One particularly useful application of satellite imagery in assessing NWP forecasts is in the case of a frontal band of precipitation occurring at temperatures which are borderline between rain and snow at the surface. In such a case the precipitation is usually forming within the clouds as ice and may melt into rain before reaching the ground if the subcloud layer is warm enough. If the precipitation is heavier than predicted by the model, the evaporation below the cloud base may be enough to cool the subcloud layer sufficiently that the precipitation reaches the ground as snow rather than rain. Errors in the forecast precipitation type are particularly obvious to the general public and will tend to attract particular criticism. Figure 6.1a shows a precipitation forecast from the UK Met Office NWP system for the afternoon of 30 January 2003. The diameter of each circle is proportional to precipitation rate, the open circles show rain

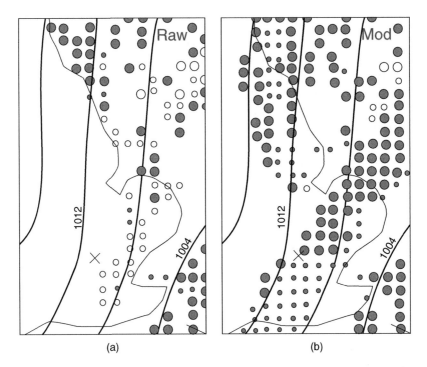

Figure 6.1 Precipitation rate forecasts from the UK Met Office NWP system for late afternoon of 30 January 2003. Circles indicate grid points with precipitation, with circle diameter being proportional to precipitation rate. Open circles indicate rain and filled circles indicate snow. (a) Raw output from the NWP model; (b) the forecast after manual modification by a senior forecaster. (© Crown Copyright 2005, Met Office.)

at the surface and the filled circles indicate snow at the surface. Using the rainfall radar, the duty forecaster at the time noticed that rain rates in the frontal band were higher than forecast. This, combined with the forecaster's knowledge of the systematic difficulties of this particular model in predicting the transition from rain to snow in frontal bands, led the forecaster to modify the forecast to include heavier precipitation rates and more snow, as shown in Figure 6.1b. During the late afternoon the rain did indeed turn increasingly to snow, leading to very significant accumulations across eastern England.

This example brings us to the following question: having spotted differences between an NWP forecast in its early stages and the true state of the atmosphere, how should the discrepancy be dealt with? One option which is out of the question is to re-run the entire forecast on the supercomputer, using new observational data to initialise the new forecast. There are several reasons why this is impractical. Firstly re-running the forecast would take too long when there are customers who need forecast information with very strict deadlines. Secondly, the supercomputer is almost certainly already being used for another purpose. The operational schedule of the UK Met Office NWP system shown in Table 5.1 shows that there is almost no time when the computer is not in use. Thirdly; there may not *be* any new observations which could be included in the data assimilation that would influence the forecast and, even if there were, there is no guarantee that they would have enough influence on the forecast to make a significant difference.

Instead, a number of more pragmatic possibilities exist to deal with model errors in the early stages of the forecast:

- The senior forecaster, or forecasting team, could simply issue text guidance to other forecasters in the forecast production chain, explaining the problem with the model forecast and suggesting how it should be corrected. Those forecasters can then decide how to incorporate the guidance into the forecasts that they issue.
- Specific changes to the model output could be made using a computerised system which allows grid point values to be modified by a senior forecaster. This might be done, for instance, by selecting a geographical area within the model domain and a period within the forecast and then applying a multiplying factor to, say, precipitation rate or temperature within that area and period. This type of modification would also be accompanied by text guidance to other forecasters explaining the reason for any changes to the raw model output.
- Dynamically consistent changes to model output could be made using a computerised system. For instance, if a senior forecaster wanted to adjust the position or central pressure of a depression, this could be done via a computer interface which would then systematically adjust all the other model generated fields (winds, temperature etc.) to be dynamically consistent with the change made by the forecaster. This change could then be propagated through the entire period of the forecast or through just part of the forecast, gradually ramping down until, at some specified point in the forecast period, the forecast returns to its unmodified version.

Clearly, the first of these options is the simplest and the third is actually very complex indeed. However, since 1997 such a system has been operational at, for example, the UK Met Office, allowing complex modification of NWP forecast model output by a forecaster using a graphical interface. Other Met Services are now also using similar systems. This type of system is described in more detail in Section 6.1.3.

Senior forecasters have been issuing text guidance to other forecasters in their organisations for many years, discussing the main thrust of the NWP forecast and any potential problems with the model prediction. This type of guidance was particularly crucial in the period when outstation forecasters (i.e. those at a location remote from the main forecasting centre, such as an airfield) had very limited access to NWP output. These forecasters could then use the text guidance to modify their own forecasts, ensuring a degree of consistency across the forecasts issued by the forecasting organisation. Today it is very much the case that output from NWP is used to generate many forecast products directly – for example graphics for television, text and graphics for internet-based forecasts or numerical forecasts for customers, such as gas and electricity providers, who need hourly predictions of weather variables such as temperature to feed into their own demand models. It makes good sense, therefore, for any modifications to raw NWP output to be made centrally by a senior forecaster (or forecasting team) rather than on an *ad hoc* basis by many different forecasters across the organisation. In this way consistency is ensured and forecaster's time is saved; the time can be more usefully used in creating forecast products for customers rather than checking and modifying the raw NWP output fields. Mylne and Grahame (2011) describe a system where, given the requirement for automation, an alternative ensemble model member which is more faithfully simulating the early observations can be selected, in place of the deterministic solution, to power forecast system requirements.

6.1.2   *Medium-range assessment of NWP forecasts*

For the short range forecasters can use observational data to check the quality of NWP output and modify it if necessary. Beyond this time range there are clearly no observations of weather that have not yet happened, so one might think that the raw output from the NWP system must be used as it stands. However, it is very much the case that senior forecasting teams are also assessing the NWP forecast beyond 24 hours and will sometimes modify the NWP output. In this case, the motivation to modify the model output comes not from observations but from other NWP models. At all the major forecasting centres, as well as using output from the 'in-house' NWP system, forecasts from several other NWP models (and in particular global models) will almost certainly be available. For instance, most European Met Services receive forecasts from ECMWF as well as their own NWP model, and output from the NWP models run at NCEP in the United States is freely available to anyone. In addition, most major Met Services agree to share their model output, so on any given day a senior forecaster may have forecasts from perhaps five different NWP models, which may include ensemble as well as deterministic forecasts. With today's high bandwidth communication networks, output from a global NWP model run at a forecast centre on the other side of the world can be available to a senior forecaster almost as quickly as the output from the model run within the same building.

Decisions on whether to modify the output from an NWP system on the medium-range time frame will be based on comparison with other NWP models. Senior forecasters will have a good working knowledge of the performance characteristics of all the models available to them. In fact, regular statistics of the performance of all the world's global NWP models are regularly published (Chapter 8 provides more details). As well as considering the mean skill of each particular model, forecasters will know the particular synoptic situations in which their own model performs well or less well.

A senior forecaster assessing the output from an NWP model on the medium range will usually look to see how close to other NWP models their own model is on 48–96 hour lead times. If their own model is a distinct outlier then there may be grounds for modifying the forecast to bring it more in line with the overall consensus. However, it may be that the prevailing synoptic situation is one in which the in-house model is known to perform very well compared with other models, so the motivation to modify the forecast output is weakened. In general, the ECMWF global deterministic forecast is statistically the most accurate global forecast model in operational use, so if there are large differences between a particular NWP model and the ECMWF forecast then the inclination may be to follow the ECMWF model. However, because ECMWF runs an ensemble forecast with 51 members, it is also possible to check whether the ECMWF deterministic forecast is an outlier within the ensemble which would introduce a note of caution. Another factor that a senior forecaster will be looking at is the evolution of the atmosphere predicted by previous runs of the NWP models. In more unpredictable situations the medium-range forecast may oscillate very wildly between subsequent runs, indicating a high degree of sensitivity of the forecast to the initial conditions in that model. In such a case the forecaster might be inclined to place more trust in a forecast from a model which is not showing radically different forecasts in subsequent forecast runs, avoiding so-called 'jumpy' forecasts.

6.1.3   Forecast modification tools

The modification of NWP output by human forecasters is now a widespread practice in major forecasting centres. At the National Meteorological Centre of the UK Met Office, senior forecasters have had a tool to do this since 1997 and NWP output from the suite of models is often modified prior to issue to production forecasters. More recently the Australian Bureau of Meteorology has been introducing a system whereby forecasters at its regional forecast centres modify the output of their numerical models before this output is used in the automated production of text and graphical forecasts. It should be pointed out that many of the changes to raw model output introduced by forecasters are small and the January 2003 example shown in Figure 6.1 is rather unusual in its radical changes.

The tools available to forecasters to modify NWP output can be very sophisticated and are designed to introduce changes to the forecast which are consistent across all the forecast variables. If a forecaster decides for instance that the centre of a depression needs to be moved or deepened all the other fields associated with that depression (winds, temperature, clouds, precipitation etc.) need to be modified in a consistent way. The UK Met Office On-screen Field Modification (OSFM) tool uses potential vorticity inversion to maintain this consistency. The forecaster uses their computer mouse to select a region in which they want to modify the raw forecast output. They can then either drag that region around the screen to move it or use a user interface to change pressure values within the selected region. The tool then uses potential vorticity inversion, together with a simple set of balance assumptions, to modify all the other fields (geopotential height, temperature, winds etc.) to be consistent with this change in all three spatial dimensions. Any changes can also be time linked throughout the forecast to prevent sudden changes at the point at which the modification is made. For instance, if a forecaster decides to modify the surface pressure field from the model at T+48 in order to bring it more in line with forecasts from other forecasting centres, they can time link the changes back to any prior time in the forecast (say T+24) so that the changes to the raw NWP output ramp up gradually from zero at this time to a maximum at T+48. The changes can also be linked forward in time so that they gradually ramp down, again until some specified point in the forecast (say T+96), or, alternatively, they can be carried consistently right through the rest of the forecast period. Because the full forecast equations do not need to be re-integrated this method is almost instantaneous. Carroll (1997) provides more details of the methodology employed by this tool.

Figure 6.2 shows two screen-shots from the UK Met Office OSFM tool. Figure 6.2a shows the pmsl, 850 hPa wet-bulb potential temperature (θ_w) and precipitation fields from the T+0 frame of an NWP forecast. The dotted red circle shows a depression centre which the forecaster has decided to move slightly towards the north-east on the basis of satellite imagery. Figure 6.2b shows the resulting T+0 frame after this movement has been made. Note that the frontal rain bands associated with the depression have also moved and the anticyclone to the south-west of the United Kingdom has become more circular in shape. Note also that the fields over Europe and the Mediterranean are unchanged.

The example shown in Figure 6.2 is fairly typical of the sort of modifications that might be made to NWP output from the early stages of a forecast (in this case at T+0). Changes at this stage tend to involve *small* shifts of particular

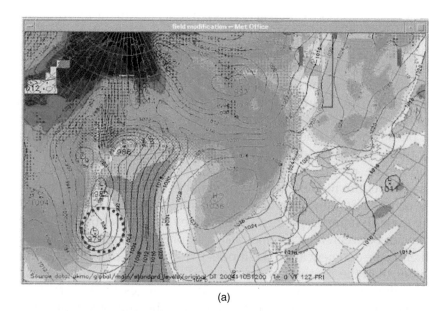

(a)

(b)

Figure 6.2   Screenshots from the UK Met Office On-Screen Field Modification tool. Black contours are mean sea level pressure, colours are 850 hPa wet-bulb potential temperature and circles are precipitation (with diameter proportional to precipitation rate). (a) shows the raw NWP output with the red dotted circle highlighting a depression centre that the forecaster wishes to move. (b) shows the modified output after this depression centre has been shifted slightly to the north-east. (© Crown Copyright 2004, Met Office.)

features or *small* changes to intensity. This is a reflection of the fact that modern data assimilation methods, coupled with a wide variety of observations from different platforms and particularly from satellites, generally do a good job of setting the initial conditions for a forecast. Changes to the raw forecast output at longer time ranges (between T+48 and T+96, say) can often be much larger than this example as the senior forecaster may want to introduce significant changes to bring the forecast more in line with that from other centres, particularly if other models are known to perform more skilfully in the prevailing synoptic conditions.

At short time ranges, a range of options which are less radical than modifying the entire model output are available to the forecaster. Particularly in cases of precipitation over land, there may be nothing wrong with the synoptic scale dynamical pattern predicted by the model but there may be discrepancies between forecast and observed precipitation. In such a case it is more likely to be deficiencies in the model parametrizations which lead to the error rather than the resolved dynamics and, in such cases, it is appropriate to modify the precipitation without making changes to the pressure and wind fields. A forecaster may choose to multiply the intensity of precipitation over a specified region by a weighting factor to either increase or decrease precipitation rates without changing the overall distribution. Alternatively, they may wish to increase or decrease the density of showers within a particular region without changing the intensity, or introduce showers into an area previously unaffected by them in the original model run. Changes of this type will be generally made in the early stages of the forecast (when radar information is available to verify the forecast) and are set to ramp down towards the raw model forecast over a period of less than 24 hours.

With sophisticated tools which allow the NWP output to be modified, forecasters must resist the temptation to 'play God' with the forecast. Any modifications must be carefully considered and be backed up by solid evidence, such as radar/satellite imagery, in the early stages of the forecast or predictions from other forecast models in the longer time ranges. Forecasting centres will periodically perform verification exercises on the raw and modified versions of the NWP output to check that the senior forecasters are not actually degrading the quality of the forecasts overall by modifying the output. It is expected that as numerical models, observations and data assimilation schemes become more sophisticated, the skill of the raw model forecasts should continue to increase, which should lead to fewer and less radical interventions by forecasters as time goes on.

A clear advantage of having a team of senior forecasters modifying the NWP forecasts at source is that all the other forecasters in the organisation receive a set of forecast data that is consistent across the whole organisation and which is ready to use without further modification. Because so many forecasts are generated directly from the numerical output of NWP models by automated computer programmes, it makes sense that any modifications to the output are made once only at the highest level in the organisation and then communicated downwards rather than being made on an *ad hoc* basis by many forecasters who may not be communicating directly with each other.

All modifications to the raw model output will generally be accompanied by some text guidance from the senior forecasters explaining what changes they have made and why. This text can then be read by forecasters who are producing forecasters for customers. In the case of guidance on the forecasts run by the US National Weather Service, this guidance is issued in the form of a web page, so that anyone can read it. This text may also be reinforced by face-to-face or telephone briefings. Another important element of guidance issued by the senior forecasting team is an indication of confidence in the forecast. With the widespread use of ensemble products this confidence may even be given in quantitative terms. In some cases, senior forecasters may even discuss alternative forecast scenarios if there is a significant probability of alternative developments. Again, ensemble forecasts are at the heart of such considerations.

6.1.4   Forecasting between the NWP forecasts

Much of the work of the senior forecasting team within a National Weather Service is involved with the assessment and potential modification of the forecasts produced by the suite of NWP models run by that Service. Particularly at centres that run more than one type of model there is a continuous cycle of one model starting to run as another finishes, so there is very little time left between assessing forecasts from different model configurations. However, the senior forecasters will also take time to monitor the developing weather situation, both locally and also on a wider scale. This wider viewpoint is very important so that forecasters are not caught out by rapidly evolving developments initially outside their immediate area of interest. Satellite imagery is a key tool for this type of monitoring and with the current generation of geostationary satellites now producing imagery which updates every 15 minutes forecasters have a good supply of up-to-date material with which to perform this task.

One very important job of the senior forecasting team in any National Weather Service is the issue of warnings of severe weather. Of course, what constitutes severe weather varies considerably around the globe, but most National Weather Services will have a responsibility to warn the public, government and other relevant agencies of any significant weather risks. The main tool used for this work will be, in the first instance, NWP output. Ensemble products can also be used to quantify the risk of a particular

Weather anomalies predicted by EPS: 20100226 00 UTC
1000 hPa Z ensemble mean VT: Sunday 28 February 2010 12UTC
and EFI values for 24h Total precipitation, 10m wind gust and 2m temperature
VT: Sunday 28 February 2010 00UTC - Monday 01 March 2010 00UTC

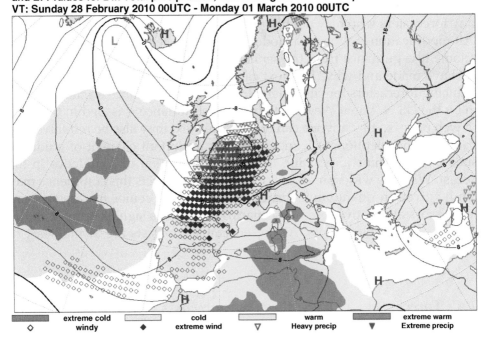

| extreme cold | cold | warm | extreme warm |
| windy | extreme wind | ▽ Heavy precip | ▼ Extreme precip |

Figure 6.3 Weather anomalies predicted by the 20100226 00UTC run of the ECMWF ensemble prediction system. Black contours are ensemble mean 1000 hPa geopotential height valid for Sunday 28 February 2010 12UTC. Extreme Forecasting Index (EFI) values for 24h Total precipitation, 10m wind gust and 2m temperature are valid for the period Sunday 28 February 2010 00UTC – Monday 01 March 2010 00UTC. (Reproduced by permission of ECMWF.)

event occurring. For example, Figure 6.3 shows some example output from ECMWF's own 'Extreme Forecasting Index' (EFI), the value of which is dependent on both the number of ensemble members simulating severe weather and the severity of the forecast relative to the model climatology. Forecasters will also be monitoring weather observations across their area of interest, checking whether warnings need to be issued in response to the observed conditions approaching particular warning thresholds, and whether warnings that are already in force need to be modified, extended or cancelled. Because weather warnings are issued for unusual or extreme events it is sometimes the case that NWP forecasts can fail to predict their occurrence. Forecasters will have a good knowledge of the tendency of their particular NWP model to under- or indeed over-predict the occurrence of specific extreme events and they will use this knowledge to guide them when deciding to issue warnings on the basis of each new forecast.

In some cases, such as the occurrence of tornadoes for instance, NWP models are not able to forecast the actual phenomena for which warnings need to be issued. Even NWP models with grid boxes on the scale of a kilometre cannot explicitly represent tornadoes, although they can indicate the probability of intense convection, which is a pre-requisite for tornado formation. In these circumstances general warnings for the probability of tornadoes can be issued 24 hours or more in advance of a potential outbreak on the basis of NWP models. However, the issue of specific warnings of location and intensity is definitely a *nowcasting* problem, with forecasters using satellite imagery and, in particular, Doppler radar to identify the parent convective cells. Since Doppler radars generally cannot identify actual tornadoes, even this methodology is highly uncertain and often the first indication of a tornado is when it actually touches down and starts to cause damage.

6.2 Production of forecasts for customers

Forecasts issued to customers take many forms, are issued via a wide range of media and may be bespoke (customer specific) or generic. The space and/or time devoted to the forecast also varies significantly from a ten second national forecast broadcast on the radio to a detailed bespoke web-based forecast provided for a commercial customer, such as wind at multiple flight levels and significant weather conditions, for multiple domains, for aviation users. Regardless of the type of user of a weather forecast service, one thing that is always true is that the forecast should be provided using language which is unambiguous, easy to digest and suited to supporting the user's decision making.

6.2.1 Broadcast and print media

Arguably the most familiar forecasts are those we hear on the radio and see on the television, procured and presented on our behalf by broadcasting organisations. These forecasts must precisely fit the broadcasting schedules, although it is often now possible to listen or view such broadcasts again online. A particular challenge with radio forecasts is the clarity with which areas affected by different weather types are subdivided in order that the listener may easily recognise the information which is relevant to them. Precise yet jargon-free language is also needed, making good use of easily recognisable place names and topographic features. Radio remains an especially important means by which forecasts can be broadcast for the benefit of shipping, although mobile phone and satellite-internet are also becoming increasingly important for such forecast users.

Television forecasts make heavy use of weather symbol maps, powered by NWP model data feeds, supplemented by presenter commentary, to get

Figure 6.4   A modern-day TV weather presentation studio. In this case the graphic on-screen is a satellite image and rainfall radar composite showing how weather conditions had evolved earlier in the day. (Photo reproduced courtesy of Weatherquest/BBC Norwich Look East.)

the forecast message across (Figure 6.4). Less frequently today, synoptic weather systems are also shown via surface pressure charts. Animated radar imagery is used extensively to demonstrate how rainfall intensity has varied in the preceding hours and model simulations are also used, in a stylised form, to extend equivalent simulated information into the forecast period. The position and behaviour of the polar front jet stream is used extensively in some countries to help explain the path and development of key weather systems. Probability forecasts, based on ensemble model output, are beginning to make inroads into television forecasts (Figure 6.5). This type of forecasting has been common in the USA for many years but is less common in Europe.

Weather forecasts remain a common sight in daily hard copy newspapers (and on their associated web sites), even though such forecasts may be 12+ hours old by the time readers see them. Arguably the ancillary information accompanying the newspaper weather forecast, such as the previous day's observed weather conditions around the world, tide times around the coast and moon phase (Figure 6.6) are of comparable interest and value to readers who may actually already have seen a more up to date forecast via another media outlet.

Recent weather and the weather forecast itself either at home or abroad do, from time to time, also become the main news item in the media. At such times, extra broadcasts and interviews with forecasters may be requested and successful communication becomes particularly critical. Unfortunately, it is commonplace that the accuracy of the information is lost as the story

Figure 6.5 Sample TV Probability Forecast. (Reproduced with permission from "The role of the broadcaster in communicating forecast uncertainty". Presentation at European Meteorological Conference 2010 by Jay Trobek, KELO-TV, Sioux Falls, South Dakota, USA; http://presentations.copernicus.org/EMS2011-798_presentation.pdf.)

gets hyped. Indeed, sections of the media often seem keen to highlight weather risks even when the provenance of the advice is dubious. One common problem is when wind-chill temperatures (the perceived 'feel' resulting from the combination of temperature, humidity and wind) are confused with the air temperature itself. Confusion can often result as to whether records have been broken.

6.2.2   Agriculture

Farming has long been a consumer of weather forecasts for the planning of the many seasonal tasks on the land. While farming forecasts include familiar surface weather variables like wind speed, to aid with the planning of spraying jobs, and rainfall to highlight irrigation requirements (Figure 6.7), sector-specific needs include estimates of evaporation and the feeding of forecast weather variables into disease prediction models which are typically used by agronomists. Steady modernisation in farming means that today forecast dissemination is required across the full range of platforms, from traditional fax and speak-to-a-forecaster phone services, through e-mail, web and mobile phone, as precision farming demands more and more information access direct to the cab of large scale machinery. With farm profitability increasingly linked to global food and energy prices, and supply contracts being signed months or even further ahead, larger-scale farmers in

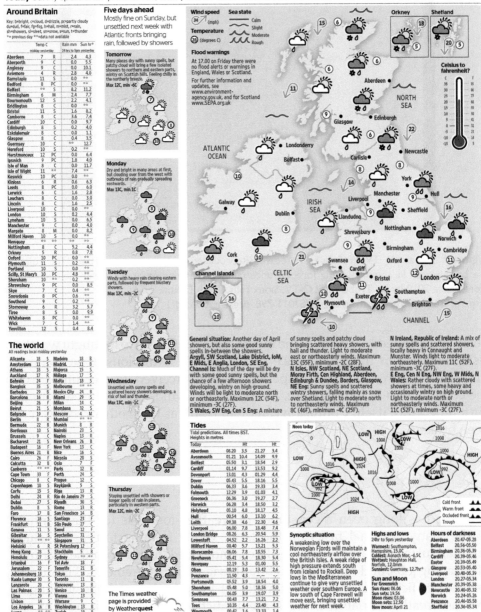

Figure 6.6   Daily weather panel from the Times Newspaper, UK. (Reproduced by permission of The Times / NI Syndication.)

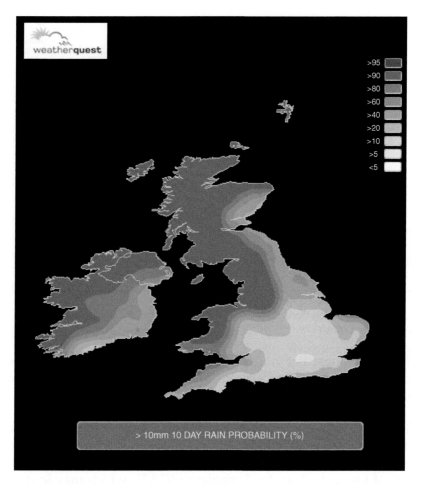

Figure 6.7 Probability of >10 mm of rain in the United Kingdom over the next 10 days (based on the ECMWF ensemble prediction system).

the developed world are increasingly likely to have an eye on the forecast for other major growing areas as they are on the local forecast. Indeed, there is a growing appetite in farming for longer-range monthly and seasonal forecasts in geographical areas where associated forecast skill is being demonstrated.

6.2.3   Civil aviation

We saw in Chapter 3 that commercial aviation is now providing a platform for making and sharing weather measurements, in the form of both vertical profiles during take-off and landing and of cruising altitude conditions in the upper troposphere and lower stratosphere, at least along

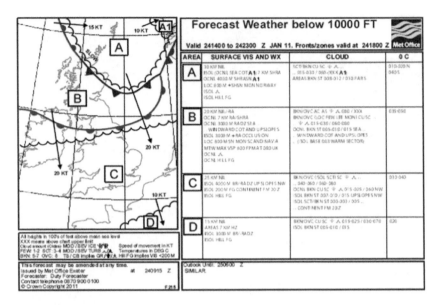

Figure 6.8 An aviation low level significant weather chart for the United Kingdom. (© Crown Copyright 2012, Met Office.)

well-trafficked routes. For much longer though, commercial aviation has been a consumer of weather forecasts. The most critical forecast information required includes:

- Winds at flight levels to help with route planning which maximises tail winds, minimises head winds and manages fuel requirements – strong winds around the jet stream will be of particular interest (Figure 6.9).
- Regions of forecast Clear Air Turbulence (CAT), which may lead to dangerous or uncomfortable flying conditions.
- Icing risk to assist with the identification of de-icing requirements before take-off.
- Areas of 'Significant Weather' which should be avoided en route or which may affect take-off or landing, perhaps causing delays or the need for diversion, with resulting extra fuel requirements and airport charges (Figure 6.8).
- Details of the expected weather conditions at the destination airfield and also for a range of alternative airfields in the vicinity if the flight needs to be diverted for some reason. These forecasts are provided in the form of Terminal Airfield Forecasts (TAFs) which follow an internationally recognised, semi-coded format and follow a set of rules for their composition which are laid down by the International Civil Aviation Organization (ICAO).

Civil aviation is a good example of the 'man–machine mix' of forecast production. Many of the forecast products provided to commercial aviation can be generated automatically from the output of an NWP model. For instance, maps of wind strengths and directions at different flight levels can be taken directly from NWP models. The UK Met Office and the US National Weather Service are designated by ICAO as World Area Forecast Centres (WAFCs) with responsibility to generate wind and significant weather forecasts for commercial aviation anywhere in the world, much of which is done automatically from NWP output. However, significant weather charts, such as the example shown in Figure 6.8 require a high level of human input, based on interpretation of the available NWP forecasts.

TAFs can be generated directly from raw or modified NWP output. Because these forecasts follow a very strict set of rules for their composition, it is easy to write a computer programme that applies those rules to NWP output to generate a forecast.

The example below shows the TAF for London Heathrow Airport in the United Kingdom issued on 13 April 2012, and covering the period between 12:00 UTC on 13 April and 18:00 UTC on 14 April.

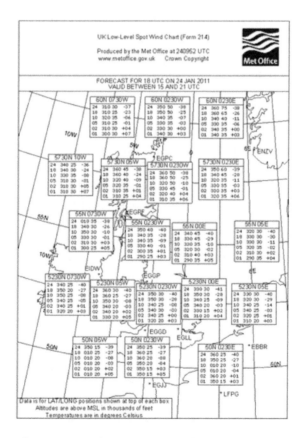

Figure 6.9   A forecast chart of winds on aviation flight levels for the United Kingdom

EGLL 1312/1418 VRB04KT 9999 SCT038 TEMPO 1312/1320 8000 SHRA
PROB30 TEMPO 1313/1319 4000 +SHRA BECMG 1400/1403 01010KT
7000 RA BKN014 PROB30 1405/1409 BKN008 BECMG 1408/1411 9999
NSW SCT020

The information content of this forecast breaks down in the following way:

EGLL is the ICAO location indicator for Heathrow;

1312/1418 is the forecast period (12:00 UTC on 13 April to 18:00 UTC on 14
April);

VRB04KT is the forecast of wind direction and strength at 10 m above ground
on the runway – in this case variable direction at 4 knots;

9999 indicates that the visibility will be in excess of 10 km;

SCT038 means that the lowest cloud layer will be scattered cloud with a base
at 3800 feet above ground;

TEMPO 1312/1320 means that the weather conditions will change temporarily between 12:00 UTC and 20:00 UTC on the 13 April to the following:
8000 is visibility of 8000 m and SHRA indicates that this will be due to rain
showers;

PROB30 TEMPO 1313/1320 means that there is a 30% probability between
13:00 UTC and 20:00 UTC of a temporary deterioration to the following
conditions: 4000 is visibility of 4000 m and +SHRA means heavy rain
showers;

BECMG 1400/1403 indicates that between 00:00 UTC and 03:00 UTC, the
weather conditions will change to become the following: wind direction
and speed of 010 degrees at 10 knots, 7000 metres visibility in rain (RA);

BKN014 indicates broken cloud with a base at 1400 feet above ground;

PROB30 TEMPO 1405/1409 BKN008 indicates that there is a 30% probability
that temporarily between 05:00 UTC and 09:00 UTC on 14 April the lowest
cloud layer will be broken cloud with a base at 800 feet above ground:

BECMG 1408/1411 9999 NSW SCT020 means that between 08:00 UTC and
11:00 UTC on 14 April the weather will change to become: visibility in
excess of 10 km, no significant weather conditions (NSW) and the lowest
cloud layer will be scattered cloud with a base at 2000 feet above ground.

Most of this content could be generated automatically using the raw or
modified NWP output for the model grid point nearest to Heathrow. The
probability statements of specific conditions would have to be added manually by a human forecaster and the whole forecast would need to be checked
by a forecaster prior to issue, but if an automated script is used to generate
the basic text of the TAF then this can save forecasters a considerable amount

of time. Given that every airfield in the world which is open to general aviation traffic must have a TAF available then this time saving is a very welcome one.

Most commercial and private civilian pilots obtain their forecast information remotely from the forecasters who produced it. Most airfields and commercial airline offices have self-briefing computer terminals. Pilots use these to obtain all the route specific information they need to plan their routes, timings and fuel requirements.

6.2.4 Military aviation

Forecast provision for military aviation is very different from provision for civil aviation with generally much more input from human forecasters. Most military aviation organisations employ their own forecasters in uniform. The UK Royal Air Force is an exception to this, with forecasters belonging to the civilian Meteorological Office. Similar forecast products to those available to civil aviation can also be used by the military. Military forecasters produce significant weather charts and TAFs in a similar way to civil forecasters. However, the nature of military aviation is such that it is often much more weather sensitive. Low-level, high speed missions by fast jets or battlefield operations and training exercises by helicopters are both very sensitive to the weather in the boundary layer and are, therefore, much more susceptible to local weather variations. Most military airfields (or aircraft-carrying ships) have their own team of forecasters working on site. These forecasters will brief the aircrew several times a day on the general weather conditions expected in the area of operations. They will also be able to provide route forecasts and specific details for particular operations. Because of the weather sensitive nature of many military aviation operations, it may often be advice from the forecaster which makes the final decision on whether a mission goes ahead or not. The close relationship between military aviators and forecasters also means that the forecasters can get rapid feedback on their forecasts and develop a good working knowledge of the types of conditions that are of most importance to particular types of mission.

6.2.5 Energy generation

Both the demand for electricity and the generation efficiency of it, using conventional turbine-based thermal power stations and many renewable technologies (e.g. wind, solar and wave), are very sensitive to weather conditions. Electricity generating companies will firstly want to be aware of the fluctuations in electricity demand since this affects price. Weather affects demand for both heating and lighting and can also create hazards for electricity distribution, especially in periods of ice and freezing rain risk for electricity pylons. Turbine efficiencies, meanwhile, are sensitive to fluctuations in atmospheric pressure, temperature and humidity and intermittency in renewable energy resource strongly affects the output from weather-dependent renewable technologies. In electricity markets where, for example, contracts are signed a day ahead, errors in the forecast and, therefore, in electricity output prediction, can lead to significant financial losses (Parks, 2011).

Accurate forecasts for electricity production installations are, therefore, very important. The accuracy of forecasts for such sites can be enhanced through the generation of site-specific forecasts which include post-processing of the raw forecasts using a Model Output Statistics (MOS) approach, made possible through the recording of on-site weather measurements. Such approaches can become quite sophisticated, varying according to time of year, wind direction (many such plants are located on the coast) and synoptic situation.

The largest developments in wind energy are generally now concentrated offshore, using 100 m+ hub-height turbines. To service this, 100 m mean wind speed is now a standard output variable from, for example, the ECMWF model, explicitly accounting for the impact of varying atmospheric stability on wind speed, rather than requiring a height correction factor which assumes a logarithmic wind profile and neutral stability. Wind and wind power output forecasts have thereby improved significantly over recent years. The trend towards coupled NWP modelling, resulting in the availability of parallel wave forecasts (Section 5.4.1), is an important development which assists with the scheduling optimisation of offshore renewable energy infrastructure servicing. By contrast, solar power forecasting is still in relative infancy and an active area of research, as forecasters look to harness the combined power of short-range forecasts based on extrapolation of satellite-sensed cloud conditions and longer-range predictions dependent on simplified NWP cloud parameterizations. The intermittent, and only partly predictable, output from installed renewable energy plant presents a particular challenge for electricity grid operators who are tasked with keeping demand and supply in close balance.

Monthly and seasonal forecasting is also of interest to the energy sector. First of all, advanced warning of the likely demand for gas helps to ensure adequate supplies, much of which may have to be imported. Another example is that hydroelectric capacity can be managed, through the geographical re-distribution of water resource, given appropriate pumping infrastructure, if dictated by longer-range skillful precipitation forecasts (Berthelot *et al.*, 2011). Forecasts even form the basis of operational weather modification techniques which can lead to enhanced winter snowfall in the mountains, later providing enhanced snow melt and water resource for hydroelectric power production in the spring (Manton *et al.*, 2011; Manton and Warren, 2011).

6.2.6   Road transport

Forecasting for road travel can take a number of forms. Firstly, in relation to road maintenance, we should consider both roadworks (repairs) and gritting in wintry weather. Successful repairs to road surfacing, involving the setting of ashphalt, require suitable weather conditions; repair teams also need to be in the right place at the right time to be able to progress a portfolio of maintenance jobs. Meanwhile, the gritting of roads in conditions of ice risk is a particularly high profile and safety-related activity, likely to attract negative publicity if the risk is mis-judged. Even if the forecast is correct in its simulation of the weather evolution, salt applied too early can be washed from roads by rain which precedes snow. Heavy traffic, responding to a forecast of impending snowfall, can itself make it more difficult for gritters to reach the areas which need to be treated in time. Despite such difficulties, forecasts for road gritting now benefit from the development of sophisticated ice risk models, which depend upon detailed energy balance assessments applied to different types of road surface. Increasingly, thermal imaging is used, across a road network, to characterise these spatially variable thermal characteristics, helping to highlight stretches of road which are most at risk through a form of post-processing of the raw model forecast. These site-specific assessments also help to protect local water courses from the impact of unnecessary applications of salt. At the end of a gritting season, it would commonly be the case that the forecast provider would be expected to summarise forecast performance through a verification report comparing forecasts against roadside automatic weather station measurements. Forecasts for road gritting are a good example of a service type which needs to be available on a 24-hour basis during the road gritting season, in some cases with additional telephone support, and a reminder that shift-working remains an essential component of modern-day forecasting.

A distinct additional hazard for road transport is strong winds, especially in exposed parts of the road network. Bridges and exposed high-level routes may need to be closed either to all or at least to high-sided vehicles if critical thresholds are exceeded in the forecast. In practice, real-time roadside measurements will also be used in the decision-making process. Even when roads stay open, the closure of ports on safety grounds, in response to a forecast of hazardous wind conditions, can have significant knock-on impacts for the management of road freight; plans may need to be implemented for the temporary management of vehicles on key routes until the port re-opens.

In such cases, the ports, the public, the police and road network operators may all become users of the same basic forecast service.

6.2.7 *Sport*

Roofs on major stadia, under-soil heating and 'all-weather' horseracing tracks are practical responses which help safeguard a sporting event against the elements. Here we will briefly focus on two other popular sports where skillful short-term forecasts or nowcasts may provide a particular advantage to participants: competitive sailing and Formula One racing.

Wind conditions are, of course, a main focus for competitive sailing and digital systems which seamlessly integrate NWP model data feeds into chart plotting software for passage planning are now common. While general synoptic scale wind forecasts are widely available, the localised effect of, for example, smaller scale sea breeze systems is an important detail of significant practical application. Since the establishment of a sea breeze circulation is dependent upon land–sea temperature differences and on the relative orientations of gradient windflow and complex coastlines, mesoscale NWP models are required operating at relatively high resolution and with regular assimilation of sea temperature updates. While sea breeze effects are most pronounced close to shore, their effects can also still be detected well offshore in otherwise light wind conditions. Knowledge of the likely presence of such local effects provides both a potential edge in competitive settings and a helping hand to those simply sailing for fun. Sailing teams at international regattas often employ their own dedicated forecasters. These forecasters often make reconnaisance visits to regatta locations months in advance of a major competitive event. They will compare the weather *in situ* with the output from available NWP models to assess any systematic errors in the forecast models and identify any local conditions, such as sea breezes and wind funnelling around coastal features, which their sailors may be able to account for in their race strategies.

In Formula One racing, meanwhile, a timely nowcast of the threat of rain, allowing a team to make an optimum tyre choice or schedule a critical tyre change, can make the difference between winning and losing. Dedicated high-resolution NWP modelling, in the area of the race, coupled with close attention to local radar rainfall output, are therefore de rigeur during practice, qualifying and the race itself. Here, too, many teams employ their own dedicated meteorologists to provide on-the-spot guidance.

Aside from sporting events on specific dates, weather forecasts are also critical for major sporting championships lasting over several weeks. Organisers of such events, for example the Olympic Games, require contingency

plans in the event of inclement weather. Will the opening and closing cere-
monies be at risk, what if transport systems are brought to a standstill, are
there weather-related safety issues which need reaction? For example, at the
Vancouver 2010 Winter Olympics, a suite of NWP models and additional
observation sites were used in support of the event organisers.

Summary

- Despite advances in computing power and automation, the production
  of weather forecasts is still very much dependent on input from human
  forecasters.
- On short forecast time ranges, forecasters can use oservational data to
  verify the NWP forecast over its first few hours. This allows forecasters
  to decide how much confidence to place in the remaining period of the
  forecast.
- On longer time ranges, forecasters can use the output from ensemble
  forecast systems and from NWP models run at other forecast centres to
  assess the degree of confidence in the forecast.
- Increasingly sophisticated tools are available to forecasters to modify the
  raw output from an NWP model at source before using it to generate
  forecasts for ccustomers
- Forecasts are provided for a very wide range of customers, using a mix of
  automated forecast production and input from human forecasters.

7

Forecasting at Longer Time Ranges

In Chapter 2 the nature of the weather forecasting problem was discussed at length. The non-linear 'chaotic' nature of the atmosphere was shown to place a limit on the predictability of the atmospheric state. The upper limit on deterministic prediction is generally agreed to be around 10–14 days. The practical considerations of weather forecasting generally reduce this predictability limit to around 7–10 days, and in some particular circumstances the limit of predictability may well be considerably less than this. There are also circumstances when the limit of deterministic forecasting may actually be longer than 14 days. Particular geographical regions of the world, such as desert regions of the subtropics, experience extremely predictable weather conditions at certain times of year, making the job of forecasting the weather there somewhat easier. Even so, in these regions it still may not be possible to make detailed deterministic predictions of, for instance, the wind speed and direction or the temperature to the nearest degree beyond about 14 days.

There are plenty of customers of weather forecasts who would like information on weather conditions at longer lead times than the predictability limit. These customers may be prepared to sacrifice detailed information on exact temperatures or timing of precipitation at these longer time ranges, but would still like to have some information on the general trends expected in the weather on periods of a few weeks to a season ahead. In some circumstances, forecasts at these longer time ranges may only have marginal skill but may have considerable economic benefit for their customers. For instance, energy traders or financial institutions trading in futures of particular commercial crops may be able to make profitable decisions on the basis of a skilful monthly or seasonal weather forecast. In more general terms this type of forecast may also have real benefit for governments and populations, particularly

Operational Weather Forecasting, First Edition. Peter Inness and Steve Dorling.
© 2013 John Wiley & Sons, Ltd. Published 2013 by John Wiley & Sons, Ltd.

in countries with largely subsistence-based agriculture that is particularly vulnerable to adverse weather conditions during the growing season.

It is clear that forecasts at longer time ranges will need to be presented using a different set of terminology than shorter range forecasts. Forecasts will probably be described in terms of departures from 'normal' (i.e. climatological) conditions and will be based more on predicting general trends than specific conditions. The use of probabilities also plays a big part in monthly to seasonal prediction, helped in part by the use of ensemble forecasts to identify more or less likely scenarios.

Weather forecasting agencies and meteorological researchers devote a lot of time, effort and money on developing prediction systems to produce forecasts on the monthly to seasonal timescale. As a purely scientific problem, longer range forecasting is as exciting a challenge now as forecasting for a week ahead was several decades ago and, as discussed briefly previously, skilful forecasts at these longer time ranges can also have real socio-economic and financial benefits. In this chapter we look at where any skill in these types of forecasts can come from given the preceding discussion in Chapter 2, which suggested that predictions at these longer lead times would fail. We then go on to looking at the methods used to try to harness any available skill in producing longer range weather forecasts.

7.1   Where does the predictability come from in longer range forecasts?

To make a skilful forecast of the future state of any physical system, there must be some information within that system at the start of the forecast period which will give predictability to the system. This information must be observable and the physical mechanisms whereby the predictability from the initial information is communicated through the system during the forecast period should ideally be understood and included within the model being used to produce the forecast. The first of these conditions is a stronger constraint than the second and it is possible to make skilful forecasts based on prior information without fully understanding the physical mechanism by which the prior information affects the state of the physical system. This possibility is discussed later in this chapter. However, it is certainly the case that if we consider the physical system that we are trying to predict to be just the atmosphere *in isolation* then it is unlikely that there will be any information in the initial state of the system that will bring any predictability to forecasts on the monthly to seasonal timescale. Individual weather systems have life cycles much shorter than this, and the larger scale patterns of, for instance, ridges and troughs in the mid-latitude jet streams also evolve on shorter timescales. The atmosphere can completely change its configuration over two weeks (and sometimes much more quickly) and due to the non-linear nature of this evolution we will not be able to make skilful forecasts of the atmospheric state beyond this time barrier.

If, however, we broaden our consideration of the physical system to include other elements of the Earth–atmosphere system, then there may indeed be observable information in the initial conditions that can provide at least some degree of predictability to parts of the atmosphere. Conditions at the Earth's *surface*, both ocean and land, tend to evolve more slowly than the atmosphere itself, so if we take these elements into consideration there may indeed be some useful predictability in the initial conditions. In fact, in the case of the oceans, we should not limit ourselves to information about the ocean *surface*, as anomalies of temperature or salinity beneath the surface at the initial time of the forecast may also be able to influence the state of the atmosphere on the monthly to seasonal timescale. Ocean influence upon the atmosphere is most prevalent in the tropical regions of the Earth where high ocean temperatures and relatively stable large scale weather patterns, such as the trade winds and monsoon circulations, mean that the ocean and atmosphere are more tightly coupled than at higher latitudes. This means that, on the monthly to seasonal timescale, and sometimes even at longer lead times, the state of one is able to strongly influence the state of the other. In particular, the persistence of oceanic anomalies means that ocean conditions can affect the evolution of the atmosphere over a period of sometimes many months.

The most well documented example of an oceanic phenomenon which can influence the evolution of the atmosphere is the El Niño-Southern Oscillation (ENSO) phenomenon, whereby the state of the equatorial Pacific Ocean has an almost global influence on the state of the atmosphere – an influence which even extends outside the tropics. Almost every textbook that has ever been written on patterns of global circulation has included a figure which shows the influence of El Niño and La Niña conditions in the tropical Pacific on weather patterns right around the globe. The impact of ENSO on global weather patterns is discussed in more detail later. Whilst ENSO is almost certainly the best known example of a phenomenon that gives predictability to the atmosphere on monthly to seasonal timescales, there are many other examples, although most generally provide less reliable skill in longer range forecasts than the ENSO example, and in several cases the mechanisms which provide the link between the initial conditions and the atmospheric forecasts are still not fully understood. For instance, ocean temperatures in the North Atlantic between Greenland and Iceland in October and November have been shown to contribute to skilful forecasts of the Atlantic tropical storm season in the following Summer/Autumn but the mechanism whereby this influence occurs is not fully understood.

Because skilful monthly to seasonal prediction is largely dependent upon the presence of anomalies at the surface, this type of forecasting is sometimes described as being a *Boundary Condition Problem* (BCP), in contrast to the *Initial Value Problem* that is weather forecasting. Mathematically speaking a BCP is one in which the outcome is largely dependent on information coming in at the boundaries of the system – so in the case of atmospheric prediction this would be the Earth's surface and the top of the atmosphere. This would be true if the boundary conditions were entirely independent of the atmospheric state. However, the surface conditions such as sea surface temperature or sea ice amount over the oceans, or surface temperature and soil moisture anomalies over land, are actually often closely coupled to the atmospheric state, so it would be somewhat artificial to consider the boundary conditions to be independent of the state of the atmosphere. If we consider the system that we are trying to predict as including both the atmosphere and the Earth's surface (and possibly even the subsurface, at least over the oceans), then monthly to seasonal prediction is still an Initial Value Problem with the initial conditions extending to those of the land surface and oceans.

The various factors that can influence the evolution of the atmosphere on timescales of more than two weeks, and hence give predictability to monthly to seasonal forecasts, are listed below. The list is approximately in order of importance, although the first point on the list is thought to have at least an order of magnitude more influence on the atmosphere than the others, and the relative importance of all the others is open to debate:

1. Tropical ocean temperature and, particularly, sea surface temperature (SST) anomalies.
2. Extra-tropical ocean temperature and, particularly, SST anomalies.
3. Stratospheric anomalies and the vertical propagation of signals both up into the stratosphere from the troposphere and vice versa.
4. Extensive land surface temperature and soil moisture anomalies, particularly over large continental regions.
5. Extensive snow cover and sea ice anomalies.
6. Variability in solar output.
7. Large volcanic eruptions.

The unpredictable nature of volcanic eruptions means that, unless an eruption has already occurred just prior to the start of the forecast, then it is unlikely to help produce skilful atmospheric forecasts. In general, a large volcanic eruption, particularly in the Tropics, can inject a considerable amount of sulfate aerosol into the stratosphere where it can be rapidly spread around the globe by the strong stratospheric winds and will have residence time of a year or so. Hence, large eruptions tend to lead to a cooling of the global mean temperature for a year or so after they have occurred (e.g. El Chichón in 1982 and Pinatubo in 1991).

7.1.1 Tropical ocean temperature anomalies

The link between SST anomalies in the equatorial Pacific and anomalies in the tropical atmospheric circulation that constitutes the ENSO phenomenon has been well known for many decades and has formed the basis for a lot of seasonal prediction methods over that time. Even as far back as the 1920s the Indian Meteorological Department under Gilbert Walker was using information on the pressure patterns across the tropical Pacific to make seasonal forecasts of the strength of the Indian summer monsoon and it was Walker himself who first used the term Southern Oscillation to describe the variations in pressure between the West and Central Pacific. In the 1960s, the pressure variations of the Southern Oscillation and the SST variations of El Niño were linked by Jakob Bjerknes (1966, 1969) and since that time information on tropical Pacific SST anomalies has been incorporated into a growing number of seasonal forecast methods using both numerical models and statistical relationships between oceanic and atmospheric variables. Once an El Niño event is underway these types of forecast have shown a good degree of skill at predicting seasonal anomaly patterns throughout the tropics and also across particular regions of the extra-tropics, such as the Pacific coast of North America and the south eastern states of the USA. However, the difficulty in predicting the onset of an El Niño event prior to the spring of the year in which the event initiates means that the lead time of such forecasts is still limited.

Another problem with ENSO-based seasonal forecasts is that the relationship between El Niño and atmospheric circulation anomalies does not appear to be stationary in time. For instance, through most of the twentieth century there was a strong correlation between El Niño events and the occurrence of a weaker than normal Indian Monsoon. Since the 1990s this relationship appears to have weakened, with the strongest El Niño on record in 1997/1998 being followed by a monsoon that was close to normal. This is a particular problem if statistical forecasting methods are being used, as these methods rely on stationary correlations existing in long data records in order to form the statistical models. Statistical methods are discussed in more detail in Section 7.3.

There are many others examples of tropical SSTs having known impacts on circulation patterns and associated weather conditions. SST anomalies in the Indian Ocean are related to Indian monsoon strength and also to rainfall anomalies across east Africa. Warmer than normal SSTs in the Atlantic Hurricane Main Development Region have an effect on the number and intensity of hurricanes in the Atlantic sector. During the 2005 season, the waters of the tropical North Atlantic warmed markedly towards the end of the season, prolonging hurricane activity well into Northern Hemisphere winter. This late warming was largely due to subsurface warm anomalies surfacing during the autumn. This example emphasises the importance of considering ocean temperature anomalies below the surface when building a seasonal forecasting model.

7.1.2   *Extra-tropical ocean temperature anomalies*

Outside the tropics a much smaller proportion of atmospheric variability is explained by sea surface temperature anomalies than within the tropics. Hence, there are fewer examples of cases where atmospheric predictability can be gained from knowledge of extra-tropical SST anomalies than from tropical ones. The example mentioned above of North Atlantic SST anomalies in November being one of the predictors for forecasts of Atlantic hurricane activity the following summer/autumn is one case where a statistical correlation between the two variables is well established but the physical mechanism which links the two is not fully understood. Other research has shown that there is some small amount of skill in *hindcasts* of the state of the atmospheric North Atlantic Oscillation in winter based on ocean surface and subsurface temperature anomalies in the North Atlantic during the preceding spring and summer. However, realising that potential predictability in actual forecasts has been hard to do and internal atmospheric variability often swamps the small part of the signal due to ocean temperature variability.

7.1.3   *Stratospheric anomalies*

Stratospheric wind circulations, particularly in the winter season, are characterised by rather stable vortex circulations, centred on or near the pole – the so called *polar night jets*. In the northern hemisphere this quasi-stable set-up is punctuated through the winter season by occasional *sudden stratospheric warmings*, when the vortex is either displaced off the pole or split into smaller circulations. When these events occur they tend to be accompanied by a discernible signal in the troposphere, which can be rather persistent. Work by Baldwin and Dunkerton (2001) using a composite of 18 stratospheric warming events showed that a tropospheric signal could be discerned up to 60 days after the onset of the original stratospheric warming. This provides a source of potential predictability around the northern hemisphere mid to high latitudes – a region of the globe in which the influence of ENSO is generally considered to be rather small. Other work has shown that any influence from ENSO which may exist in mid to high latitudes may come through a stratospheric connection whereby changes in the stationary Rossby wave pattern in the Pacific allow increased vertical propagation of waves into the stratosphere which can then interfere with and weaken the polar vortex.

7.1.4   Land surface temperature and soil moisture anomalies

In certain parts of the globe there is a strong feedback between land surface conditions and local weather. In particular, persistent negative soil moisture anomalies alter the surface energy balance, increasing the proportion of incoming solar radiation that is used to raise the surface temperature at the expense of evaporating water from the surface. This has the dual effect of both warming and drying the boundary layer. A drier boundary layer may also lead to less locally generated convective precipitation because although the surface is warmer and, therefore, air parcels are more buoyant, if the rising air is very dry then convective cloud depths and, hence, precipitation amounts will be reduced.

The effect of persistent land surface anomalies will be most marked in regions with rather slowly evolving atmospheric conditions where the positive feedback between surface and boundary layer is allowed to fully develop. The effect will also be most marked in locations with a strong seasonality in precipitation, particularly in climatic regions where the majority of rainfall occurs during the winter season. In such circumstances a deficit in winter rain will lead to a soil moisture deficit at the start of the spring season, which will then in turn lead to higher than normal temperatures over the summer, as there is then little opportunity for soil moisture to be recharged. Continental interiors and semi-arid regions are examples of locations where persistent land surface anomalies can have an impact on weather patterns on the monthly to seasonal timescale. However, even places like the United Kingdom, where the weather is usually characterised by rapidly evolving synoptic scale weather systems, the impact of soil moisture anomalies is still observable, with warm summers such as 1995 and 2003 being preceded by significant soil moisture deficits through spring and early summer. Of course, in a location such as this, a spring soil moisture deficit is not by itself a guarantee of a warmer than normal summer, as large scale controls on the weather patterns will generally be more important than the local feedback between surface and atmosphere. In 2011, for instance, a very dry spring in the United Kingdom was followed by a rather cool summer.

There are also examples of situations where land surface temperature anomalies can be correlated with weather patterns at a remote location several months in advance. For example, the Indian Meteorological Department has used the surface temperature of north-west Europe during January as one of its predictors for the strength of the Indian Summer Monsoon for many years. The physical link between these two quantities is certainly not obvious and in this type of situation it may be that both European temperature and Indian Monsoon strength are actually independently correlated with a third factor, so the two may not be directly linked at all.

7.1.5   *Land surface snow cover and sea ice anomalies*

In much the same way that soil moisture anomalies can have a persistent influence on local temperatures, snow or ice cover anomalies can also affect temperature anomalies on the monthly to seasonal timescale. Particularly extensive or thick snow cover at the end of the winter season will take longer to melt going into spring, so more of the incoming solar radiation will be used to melt snow and, consequently, the land surface will heat less. The same is also true of extensive sea ice cover at the end of the winter season, so ocean temperatures will increase less quickly in the spring. As with soil moisture, the largest effects associated with this would be expected in regions with slowly evolving weather systems (e.g. regions dominated by anticyclones) to allow the feedback between the surface and the local boundary layer to have maximum effect.

Again, there are examples of cases where remote snow cover anomalies have been used to predict weather conditions at locations far removed from the snow itself. Springtime Eurasian snow cover anomalies have been used by the Indian Meteorological Department as another predictor of summer monsoon strength.

7.1.6   *Solar variability*

As the Sun is the sole source of energy for the Earth's atmosphere, it might be expected that changes in solar output would lead to changes in weather on Earth. On very long (i.e. geological) timescales it is well known that changes in the Earth's orbital parameters which affect, particularly, the seasonality of the top-of-atmosphere radiation budget lead to the formation and dissipation of ice ages. On shorter timescales, too, the reduction in solar output during the *Maunder minimum* of the seventeenth and early eighteenth centuries is thought to have been a contributing factor to the reduction in temperatures throughout Europe and parts of North America during the same period (the so-called 'little Ice Age').

A major challenge to meteorologists and solar physicists has been to pick out patterns in atmospheric circulation anomalies that might be attributable to solar variability, particularly as the effects of solar variability will often be masked by other atmospheric internal variability. Recent research has started to show a possible link between low solar activity and an increase in anticyclonic 'blocked' flow in the Eastern North Atlantic during winter (Lockwood *et al.*, 2010). This type of persistent weather pattern disrupts the North Atlantic jet stream and results in fewer Atlantic depressions travelling across Western Europe and an increase in north-easterly flow off the European continent. Whilst this is consistent with observations during the Maunder minimum and also during recent cold European winters around 2009–2011, it is clear that further work is needed in this area before this potential source of predictability can be incorporated into seasonal forecasting models.

7.2   Observations of ocean and land surface variables

Since tropical ocean surface temperatures, and the equatorial Pacific in particular, are such a crucial factor in seasonal prediction, good observations of the state of the tropical oceans are essential to be able to make use of its potential predictability. Since the 1982–1983 ENSO event, NOAA began developing and installing a network of moored buoys across the tropical Pacific to monitor surface and subsurface temperatures and currents as well as the weather conditions at the surface. This network is known as the Tropical Atmosphere–Ocean Triangle Trans Ocean Buoy Network (TAO-TRITON) array and currently consists of 68 moorings spaced across the tropical Pacific spanning the equator (Figure 7.1). Each mooring consists of an automatic weather station recording air temperature and humidity, wind speed and direction, rainfall amount and incoming shortwave radiation. Beneath the surface there is a steel cable fixed to the ocean floor with temperature sensors down to a depth of 500 m, with typically five sensors in the top 100 m.

The TAO-TRITON array provides unprecedented detail on the state of the upper ocean in the tropical Pacific and is essential for the initialisation of ocean conditions in NWP forecasts of ENSO. A similar although less comprehensive array has also been installed in the tropical Atlantic and is known as PIRATA (Prediction and Research Moored Array in the Atlantic). A similar network is also being developed in the tropical Indian Ocean.

There are still vast areas of the ocean which are not covered by this type of comprehensive moored buoy network. The massive gaps are filled to some

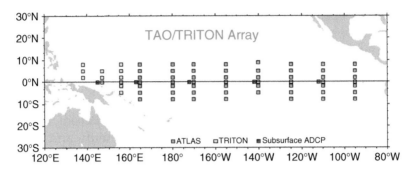

Figure 7.1 The position of moorings in the TAO-TRITON array in the tropical Pacific Ocean. (Image courtesy of NOAA PMEL/ TAO Project Office, Dr. Michael J. McPhaden, Director.)

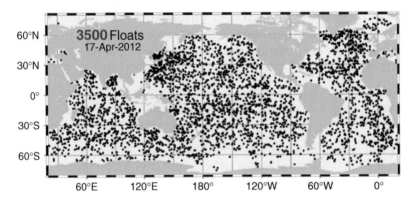

Figure 7.2 Locations of ARGO ocean temperature and salinity profiling floats on 17 April 2012. (These data were collected and made freely available by the International Argo Program and the national programs that contribute to it. (http://www.argo.ucsd.edu, http://argo.jcommops.org). The Argo Program is part of the Global Ocean Observing System.)

extent by the ARGO Float programme of drifting buoys. These buoys move around the oceans with the currents but also descend and ascend through the top 2000 m of the ocean on about a 10-day timescale, taking a vertical profile of temperature and salinity as they do so. On arriving back at the surface this profile is transmitted via satellite to a data collection centre and is then available for near real time analysis of the current state of the upper ocean. The first of these floats was deployed in 2000 and as of April 2012 there were about 3500 floats deployed and transmitting data throughout the global oceans (Figure 7.2).

Satellites are able to gather information on ocean surface temperatures with global coverage and a number of different sensors, using different wavelengths in the infrared part of the spectrum, are currently making SST observations. Satellite-based altimeters can also provide useful information on ocean surface height anomalies, which are closely related to pressure gradients and hence to ocean currents. Surface winds over the oceans are also a useful quantity in setting initial ocean conditions in NWP models; these can be estimated from satellite-borne scatterometers, which work by detecting the change in radar reflectivity due to the wind-driven ripples on the ocean surface. Of course, none of these satellite-borne techniques can see through cloud and with large parts of the global oceans being covered in clouds for much of the year, particularly in the subtropical anticyclonic regions, there are large gaps in the satellite coverage.

As far as observations of the land surface are concerned, we are again heavily reliant on satellite-based instruments. There are some *in situ* measurements of soil moisture but vast areas of the land surface have no *in situ* measurements at all. Satellite-based measurements usually rely on changes in surface radiative emissions in the microwave part of the spectrum. They are, of course, subject to large error bars, especially when the surface itself is covered in vegetation. Land surface snow and sea ice extent measurements from satellite have been made for many years but often the snow depth or ice thickness is of equal importance to the extent and this is less easy to measure remotely. Some microwave sensors have been used to measure snow depth but their operation depends on certain assumptions about the nature of the snow being measured, which may not be correct.

7.3   Monthly to seasonal forecasting systems

Having established that there is some potential predictability in forecasts on the monthly to seasonal timescale, forecasting systems need to be developed which are capable of exploiting this predictability. A pioneer of seasonal forecasting was Gilbert Walker, working as Head of the Indian Meteorological Department (IMD) in the early part of the twentieth century. At that time virtually the entire population of India depended on rain-fed subsistence agriculture and, with 80% of the country's rain falling during the monsoon season (June to September), predicting the strength of the annual monsoon was seen as crucial to the nation's well-being. Really quite small perturbations in the strength of the monsoon could have a massive impact on the country, with a weak monsoon potentially forcing many millions of people into dependence on State and National government aid. Predicting in advance the strength of the monsoon was, therefore, an obvious target for the IMD. In an age well before there were established global meteorological data sets such as the re-analysis products widely available today, Walker set a large team of staff to work looking for statistically significant correlations between atmospheric and land/ocean surface variables and the strength of the Indian monsoon. In particular, they were looking for relationships where predictor variables led the onset of the monsoon by several months. These predictor variables could then be combined into a statistical (or 'empirical') model that could be used to forecast the monsoon strength.

This type of empirical seasonal forecasting is still widely used today, despite the availability of sophisticated NWP methods. It has the advantage of being relatively cheap and easy to run, and, in some cases, particularly in the tropics, there is a strong enough relationship between the predictor variables and the variable being predicted that there is significant skill in this type of method. There are also drawbacks with this technique, too, which are discussed further in Section 7.3.1. Large forecasting centres are also investing a lot of research and development effort into NWP methods for monthly to seasonal prediction. On the basis of the preceding discussion, it is clear that numerical models being developed for this purpose will need to have some adaptations from the NWP models used for short-to-medium range deterministic forecasting. These are discussed in Section 7.3.2.

7.3.1   Empirical methods

Empirical techniques for monthly to seasonal prediction have their roots well before the formation of any meteorological agencies. The folklore of many countries is littered with examples of ways of predicting the nature of the coming season by observing signs in nature. A classic example is the English rhyme:

'Oak before Ash, we'll have a splash. Ash before Oak, we'll have a soak.'

This rhyme predicts that if the leaves on oak trees appear before those on ash trees during the spring then the following summer will be dry. If, however, the ash leaves appear first then the summer will be wet. It is fascinating to note that a similar rhyme also exists in German folklore;

'Grünt die Eiche vor der Esche

hält der Sommer große Wäsche

Grünt die Esche vor der Eiche

hält der Sommer große Bleiche.'

However, the prediction from this rhyme is precisely the *opposite* of the English version, with leaves on the oak before those on the ash predicting a summer washout! Of course, nature tends to respond to recent weather rather than acting in anticipation of future events.

It is pretty clear that relationships of this type have very little value in terms of predictability. We have to move forward to the start of the twentieth century and the pioneering work of the Indian Meteorological Department to see the start of a proper scientific basis to empirical seasonal forecasting techniques. Gilbert Walker's quest for predictors of the strength of the Indian monsoon followed on from work by two earlier heads of the IMD, H.F. Blanford and Sir John Elliot. They were initially motivated by the failure of the monsoon rains in 1877. A further severe monsoon failure occurred in 1899. In both cases the rains were deficient by more than 25% of their normal value and, as such, they rank as the two biggest monsoon failures in the instrumental record. Both led to famines, with well over a million estimated deaths in each case. Elliot established a forecasting technique based on looking at the Himalayan snow cover during March, with excess snow cover apparently leading to a weaker than normal monsoon. Weather conditions in Australia were also considered as a monsoon predictor and we now know that a summer drought on the Australian East coast is strongly correlated with El Niño events. In fact, both 1877 and 1899 were El Niño years.

Gilbert Walker set his analysis teams to work looking for correlations (which he termed *teleconnections*) which might give them a possibility of predicting the strength of the monsoon. In the course of this pursuit he discovered and defined the Southern Oscillation, a teleconnection pattern with such a vast influence that only the North Atlantic, Europe and the Arctic do not really feel its influence. We now know that the Southern Oscillation is the atmospheric response to El Niño conditions in the Eastern Tropical Pacific Ocean. Figure 7.3 shows the simultaneous correlation between surface pressure everywhere on the globe during Northern Hemisphere winter with SST in the NINO3 region in the tropical Pacific (shown by the red box). Clearly the impact of El Niño is on an almost global scale and this makes it a prime predictor for many different empirical forecasting methods since the work of Walker. If the occurrence of El Niño itself can be predicted, or if it is recognised in its early stages, then it brings a great deal of potential predictability to empirical seasonal forecasting methods.

Once a predictor, or set of predictors, has been identified by correlation analysis, a quantitative relationship can be derived between predictors and predictand through single or multiple regression analysis. This can be either linear or power regression depending on whether the relationships between predictor and predictand appear to be linear or whether there are clear non-linearities. A linear empirical forecast model takes the form:

$$R = C_0 + \sum_{i=1}^{i=N} C_i X_i$$

where R is the variable being forecast (the predictand), C_0 is a constant, the C_i terms are weighting coefficients and the X_i terms are the predictor variables.

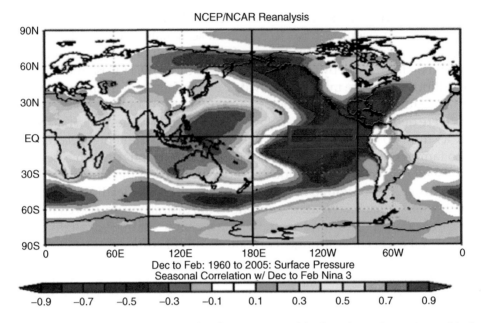

Figure 7.3 Simultaneous correlation of surface pressure in Northern hemisphere winter with the SST anomaly in the NINO3 region, shown by the red box

A model of this type is very easy and cheap to build and run. All that is needed are:

1. a long time series of the predictand variable;
2. global data sets of the predictor variables (preferably covering the same time period as the predictand time series);
3. a PC or simple computer;
4. a statistical software package.

This type of forecast model is eminently suitable for a forecasting agency or other organisation without access to the sort of high performance computing facilities used by many national meteorological services for forecasting. Predictand time series (such as national rain-gauge records) are usually held by National Met Services, the global predictor data sets such as SST records are freely available via the Internet and the statistical processing and subsequent running of such models can all be done on a laptop computer. Since these empirical methods have most potential predictability in the tropics, this means that national forecast providers in developing regions such as Africa can build and run these types of models relatively easily.

One of the best known empirical models of this type in operational use is still that used by the Indian Meteorological Department for its seasonal monsoon forecasts (although it is now also supplemented with information from NWP systems run on supercomputers). The model is regularly revised and new predictors are occasionally added whilst others are dropped. Each year the weighting coefficients for each predictor are also re-calculated. Prior to 2002, the IMD was using a 16-parameter (predictor) power regression model in which several of the predictors were related to the state of, or the trends in, the tropical Pacific SST. The skill of this model in predicting the monsoon strength had been apparently deteriorating since the early 1990s but in 2002 the model predicted a 'normal' monsoon (i.e. rainfall within 10% either way of the long period average) whereas the actual monsoon was very weak with only about 80% of the long period average rainfall. This failure prompted a radical overhaul of the method, and the 16 predictors were reduced to eight for the forecast made in spring, with two further predictors being added for an update in June.

Interestingly, hindcasts of the 2002 event made with this new forecast model have *also failed* to predict the 2002 drought. This raises one of the major drawbacks of empirical seasonal forecasting methods. That is that there may be occasions when the system under prediction is not being strongly forced by any of the predictors, and internal variability in the system is larger than any external forcing. This was apparently the case in 2002 with the monsoon drought being a consequence of an extended break in the monsoon through the middle of July when the rainfall would normally be at its heaviest. This break, in turn, was in part linked to the passage of the Madden–Julian Oscillation (MJO) through the Indian Ocean sector – an *intraseasonal* weather system with no strong response to any of the predictor variables in the IMD empirical model (Saith and Slingo, 2006).

This, in turn, highlights another drawback of empirical techniques for seasonal forecasting. If the forecast fails it may not be obvious why, so it is not always clear how to adjust the model to prevent the same thing happening again. Even in cases when the failure is reasonably well understood, such as the 2002 monsoon case, there may not be anything that can be added to the forecast model to take account of the crucial factor. There is currently no way of predicting levels of MJO activity more than a month or so in advance, either empirically or using NWP methods, and hence there is no way to include its potential impact in an empirical model.

A further drawback of empirical methods, already touched upon in Section 7.1.1, is that these methods rely on stationary correlations between predictor and predictand variables in long data records in order to be able to build the statistical models. In fact, it is rare for any correlation between two meteorological variables to be stationary over time, as there is a lot of slowly evolving variability within the earth–atmosphere system, which means that correlations wax and wane. Add to this natural variability the trends introduced by anthropogenic climate change and it becomes even less likely that correlation will remain stationary over time. This means that empirical models have to be constantly revised and re-tuned in order to keep pace with the evolving teleconnection patterns, and sometimes a model which was able to display significant forecast skill over a number of years can deteriorate in terms of its forecast skill.

The physical link between predictor and predictand variables is another issue in empirical forecasting. A couple of examples in the preceding sections have shown that although a sometimes quite strong correlation may exist, it is not always obvious what the physical link may be between the variables. In fact, simply by looking for correlations between a predictand and a number of different predictors, it is possible that spurious links are being introduced because the predictand and one or more of the predictor variables are all responding to another independent influence, which may or may not be one of the other predictors. This is particularly an issue with predictands that show a response to El Niño. Because so many aspects of the global circulation are influenced by ENSO, it may be that some of the chosen predictor variables correlate well with the predictand purely because they also correlate well with ENSO.

Despite all the disadvantages of empirical techniques discussed earlier, there are a number of operational situations in which empirical models are still at least as skilful as, or in some cases significantly more skilful than, an NWP forecast for the monthly to seasonal period. Colorado State University (CSU) is able to produce forecasts of the Atlantic Hurricane season seven months prior to the onset of the season using a very simple three parameter linear regression forecast model (Klotzbach, 2008). This version of the CSU forecasting model was developed in 2008 (although similar methods have been used at CSU for over 20 years), so as yet it is still too early to say whether it will show significant skill in real-time forecasts. It has been able to explain much of the variance in tropical storm activity from year to year when run in *hindcast* mode. No NWP methods for tropical storm prediction exist at this lead-time, so this method is widely noted by interested parties.

Because of the dominance of ENSO within the global, and particularly tropical, circulation, many empirical forecast methods include at least one ENSO-related variable within their predictor set. The Colorado State University forecasts of Atlantic Hurricane activity, for instance, use the mean sea level pressure in the centre of the tropical Pacific as one of the predictors. Clearly, if the phase of ENSO itself can be forecast in advance then it can add a great deal of potential predictability to many other empirical forecast models. It is not surprising, therefore, that a number of different methods exist for predicting some measure of the ENSO phase. Many of these rely on complex atmosphere–ocean coupled climate models, but there are also some empirical and statistical methods in use which have historically shown comparable skill in ENSO prediction to the far more complex coupled models. Some of these have moved beyond the simple linear regression-type model described earlier. Neural network methods are now being exploited to build models which are able to train themselves on past data and predict patterns of SST variability based on the observed leading modes of SST variability in the tropical Pacific. Typical predictors for statistical models used for ENSO prediction include the observed SST and wind or surface pressure anomalies across the tropical Pacific. Some models also include subsurface temperature anomalies in their predictor sets. Whilst the subsurface information can indeed add a very important dimension to the forecasting challenge, statistical models which use this information are somewhat hampered by the fact that the subsurface temperature record is rather short. A network of moored buoys with subsurface temperature sensors was installed across the tropical Pacific in response to the big and largely unpredicted 1982–1983 El Niño event, and data started to come on stream from this network during the mid-1980s (Figure 7.1). In statistical modelling terms this is still a rather short record for the establishment of statistically significant relationships between predictor and predictand variables.

7.3.2   NWP methods

Section 7.1 established that potential predictability for monthly to seasonal timescale forecasts is likely to come from knowledge of conditions at the Earth's surface, over land and, particularly, over the oceans. Hence, if NWP models are to be adapted for use in longer range forecasting it is clear that the land surface and ocean elements of the models will be key components. It is also clear that knowledge of the land surface state and the current ocean conditions, both at the surface but also at depth, will be crucial to setting the initial conditions for these components, since this is where any predictability will be found. Accurate assimilation of the ocean and land surface observational data into models for seasonal forecasting is as critical as the assimilation of atmospheric data into numerical models used for short-range weather forecasting. In this section some of the different ways that NWP models have been adapted for use in forecasting at longer time ranges are considered.

One of the first considerations of developing an NWP system for monthly to seasonal range forecasts is that the internal variability of the atmosphere in both time and space will inevitably be large at this time range compared to any local variations in weather from place to place that might be represented on a high resolution model grid. Forecasts beyond the two week predictability limit will not be given in terms of lots of local spatial and temporal detail, but will instead be describing general patterns over large areas.

Hence, NWP models used for seasonal prediction always run at lower resolution than their short-to-medium range equivalents. For instance, ECMWF runs its global *medium-range* forecast model at a spectral truncation of T1279, which is about equivalent to a horizontal grid spacing of 16 km. The ECMWF *seasonal* forecast model uses the same atmospheric model at a spectral truncation of T255, which would be equivalent to a horizontal grid spacing of about 80 km.

Whilst most NWP models used for seasonal prediction run with their *horizontal* resolution degraded from the equivalent version used for short-range forecasting, the same is not generally true for vertical resolution. Section 7.1.3 showed that vertically propagating signals both up into the stratosphere from the troposphere and also down from the stratosphere into the troposphere are potential sources of long range predictability. To realise this potential predictability it is important that NWP seasonal forecasting models have good vertical resolution, and, in particular, a well-resolved stratosphere. So, whilst ECMWF degrades the horizontal resolution of its seasonal forecasting model to T255, it retains the full 91 vertical levels of the medium range version of the model, with a model top at 0.01 hPa (approximately 75 km), which is in the mesosphere.

The next and probably most critical issue with using NWP methods for seasonal prediction is how to deal with the oceans within the model, as this is where much of the predictability will come from. There are three basic possible approaches to this question.

1. Create the best possible analysis of SST patterns at the start date of the forecast and then *persist* the SST *anomaly* pattern through the period of the forecast, using this as surface forcing to an atmosphere-only NWP model.
2. Create the best possible analysis of ocean surface and subsurface temperatures. Then use a numerical model of the oceans to predict the evolution of these temperatures through the forecast period before finally using the SST predictions from this ocean model to force an atmosphere-only NWP model. This is known as a 'two-tier' seasonal forecast system.
3. Create the best possible analysis of ocean surface and subsurface temperatures. Then use a fully coupled atmosphere–ocean NWP model to predict how the atmosphere and ocean will evolve together through the forecast period.

Option 3 on this list is the most physically realistic option, as the ocean and atmosphere are indeed a coupled system and so are able to influence each other on all time periods. However, this is also probably the most expensive option and in some cases it may be possible to justify choosing one of the other alternatives.

Persisted SST forecasts

Almost all *short-to-medium* range NWP forecasts are performed using fixed SSTs as a lower boundary condition to an atmospheric model. Forecasting centres may perform an SST analysis perhaps once a week or every 5 or 10 days, and will then use this same SST field as the lower boundary forcing for their NWP model until the analysis is next updated. This is based on the fact that, over a period of a week or so, SSTs do not change very much. If this approach is acceptable for forecasts for 10–14 days, then it is probably also acceptable for periods of 14–20 days without much detriment to the forecast. Beyond about 20 days, the seasonal cycle of SSTs may result in significant changes in SST between the start and the end of a forecast. To take account of this, forecasts of up to a month or so using an atmosphere-only model forced with SSTs at the lower boundary usually use persisted SST *anomalies*, where the SST anomalies in the original analysis are just added to the observed annual cycle of SST to create the lower boundary condition for the atmosphere model. This is much cheaper than using a full ocean model to calculate the SST at the lower boundary and in many cases, for forecast lead times of up to about one month, this method will not result in a great loss of predictability.

There will be occasions, however, when persisted SST anomalies *do* result in large discrepancies from reality, even within a forecast period of a month. For example, the El Niño of 1997–1998 underwent an extremely rapid termination in the spring of 1998. As Figure 7.4 shows, the SST on the equator at 140°W in the central Pacific fell by about 8°C in less than a month during May. A model using persisted SST anomalies through this period would clearly have failed to represent this rapid fall and, therefore, would not have captured its effects on both the local weather conditions and the large scale circulation patterns throughout the tropics. Having said that, it is also clear from this figure that monthly forecasts using persisted SST anomalies initiated any time between the late spring of 1997 and April 1998 would have included the effects of the large El Niño event that lasted through this period. It is also true that no operational coupled ocean–atmosphere forecast model captured the rapid termination of the 1997–1998 El Niño event anyway.

Two-tier forecast systems

In a two tier system, a stand-alone ocean model is used to calculate the evolution of SSTs throughout the forecast period, starting from a detailed ocean analysis. Given the dominance of the El Niño cycle on SSTs at the seasonal timescale, this ocean model may just be a model of the tropical Pacific, or it may indeed be a full global ocean model. The SSTs calculated from this model are then used as lower boundary conditions for an atmosphere-only

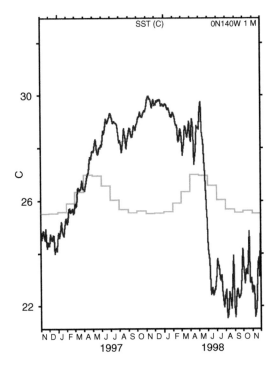

Figure 7.4   Observed SST on the equator at 140°W in the central Pacific from November 1996 to November 1998 (dark blue line). The normal annual cycle of SST at this location is shown in the light blue line. (Image courtesy of NOAA PMEL/ TAO Project Office, Dr. Michael J. McPhaden, Director.)

NWP model. In the case where SSTs are only calculated from the ocean model in the tropical Pacific, persisted SST anomalies are used across the rest of the globe.

A basic premise of a two-tier system is that, on the monthly to seasonal timescale, most of the changes in SST are due to slowly evolving ocean dynamics and are not strongly forced by the atmosphere. In terms of El Niño, many of the theories which explain its occurrence are based on slowly evolving equatorially trapped Kelvin and Rossby waves, which transmit information across the Pacific basin on timescales of several months and once they are underway they are not materially affected by atmospheric variability. Whilst this may be true to first order, it has become clear since the advent of detailed ocean observations across the tropical Pacific that each El Niño event goes through a slightly different evolution to any previous event and that much of that event-to-event variability is due to the influence of the atmosphere on the evolving ocean state. For instance, several bursts of strong westerly winds of 1–2 weeks duration during the winter and spring of 1996–1997 in the West Pacific are thought to have been instrumental in the rapid onset and large magnitude of the 1997 El Niño. Whilst the underlying ocean dynamics probably would have caused an El Niño event anyway, it may well have set in more slowly and been weaker if the wind bursts had not occurred.

Fully coupled forecasts

Neither the atmosphere nor the ocean evolves in isolation one from the other. For physical consistency, therefore, the best possible forecasting system for the monthly to seasonal timescale is one in which both the atmosphere and the ocean are able to influence and feedback upon each other. In most cases, therefore, this means that forecast centres that already have an atmospheric NWP model for short-to-medium range forecasting need to acquire or develop an ocean model and a system which will couple the two models together.

Numerical ocean models are based on the same physical principles as atmospheric models. After all, the ocean is a fluid and so obeys the laws of fluid dynamics. Thus the main principles described in Chapter 4 for an atmosphere model will also be included in an ocean model. In some ways ocean models are simpler than atmosphere models as they do not need to take into account the complex changes of phase of water that make up a crucial component of atmospheric NWP models. Where ocean models do become more complicated is in the presence of horizontal boundaries at the coastlines of ocean basins. In terms of physical parametrizations, ocean models do not have to deal with clouds, so the parametrization schemes are generally simpler than an atmospheric model. Small scale mixing of heat in ocean eddies does make a significant contribution to the ocean energy budget, so this does need to be dealt with by parametrization.

Coupling an atmosphere and an ocean model together means that relevant information needs to be passed between the two models. The atmosphere needs information from the ocean on the surface temperature and some coupled modelling systems also pass information on ocean surface currents to the atmosphere model, although this is a second order effect compared to the influence of SST. The ocean model needs information from the atmosphere on wind speed and direction and surface fluxes. Very often information is passed between the two model components once a day. This daily coupling strategy is widely used in models being used for long-term climate simulation. In models being used for monthly to seasonal prediction there are arguments that the ocean and atmosphere need to exchange information more frequently. Averaged over the whole globe and the whole year, diurnal variations in SST are rather small. However, in certain parts of the tropics, particularly those with clear skies and light winds, SSTs may vary by several degrees during a 24-hour period, in a similar way to that of land surfaces heating up during the day and cooling at night. This diurnal variability of SST can, in turn, force atmospheric convective activity which, in turn, can modify the state of the atmosphere over a period of several days to several weeks. This means that diurnal variations in SST can map onto longer term variations. Studies with the ECMWF monthly prediction system showed that, by coupling the atmosphere and ocean four times a day, forecasts of the state of the Madden–Julian Oscillation (MJO), were skilful for about five days longer than forecasts where the atmosphere and ocean were coupled only once (Woolnough *et al.*, 2007).

Monthly to seasonal predictions tend to be expressed in terms of anomalies from the normal weather conditions expected at that time of the year. As with shorter range forecasts, therefore, agencies running this type of forecast need to maintain a database of the NWP model's normal conditions, as these will certainly differ from reality. They will also be lead time dependent. This means that many years' worth of 'hindcasts' need to be run every time a new model formulation is introduced, so that 'normal' conditions for the model can be defined. Figure 7.5 shows the bias of the ECMWF seasonal forecast model for 2 m temperature in the December–February season for forecasts started in November, relative to re-analysis values. So, at a location where the model has a cold bias of 2°C, if the NWP model predicts temperatures which are 4°C warmer than the observed 'normal' conditions for this season, the actual forecast issued to customers will be for temperatures 6°C above normal. Running the hindcasts to generate the bias statistics adds more to the expense of running seasonal NWP forecasts.

This linear assumption of model bias is not necessarily correct and can lead to some problems with forecasts expressed as anomalies from normal. For instance, the atmosphere in the NWP model may respond in a different way to

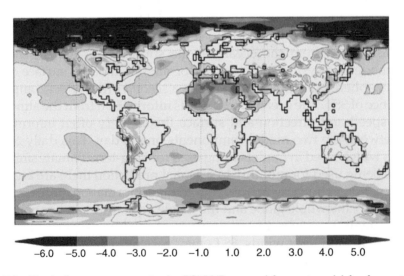

Figure 7.5   Bias in 2 m temperature in the ECMWF seasonal forecast model for forecasts of the December–February season, initiated in November. (Reproduced by permission of ECMWF.)

the real atmosphere to a particular pattern of SST anomalies, leading to model biases which are themselves dependent on specific conditions. Subtracting the long-term model bias from the forecasts will improve forecast skill when averaged over a large number of forecasts but in individual circumstances it may result in misleading forecasts. It is entirely possible, for instance, that the forecast model bias in periods when there is an El Niño occurring may look quite different to the bias when there is no El Niño. No operational NWP seasonal prediction system currently attempts to account for these potential variations in bias. This is partly because there are so many possible permutations, and even just forming bias climatologies for the simple cases of El Niño, La Niña or ENSO neutral conditions would mean that hindcasts would have to be run over much longer periods to capture enough events to populate each category. Even if this were done, every El Niño event differs in its SST anomaly pattern and the atmospheric response to that pattern, so the exercise may have limited benefit anyway.

Using a global NWP model coupled to a global ocean model has the obvious benefit of producing a forecast of any given meteorological or oceanic variable everywhere on the globe for the whole period of the forecast. This contrasts sharply with empirical techniques, which make predictions of a single variable at a fixed location for a specific period in the future. Hence, this type of forecasting is preferred by large meteorological agencies which may have customers for their forecasts based all around the globe. These large agencies are also able to afford the expense of running coupled global NWP systems for forecast lead times of several months.

Although coupled NWP models can produce forecast information all over the globe, much of that information, particularly at longer lead times, may not actually have any forecast skill because of the inherent unpredictability of the atmosphere beyond the 10–14 day predictability limit. At these timescales, forecasters are trying to isolate any predictability that may come from either the ocean or the land surface. It is only by running an *ensemble* of forecasts, therefore, that forecasters will be able to see if any signals in the forecast are indeed coming from real sources of potential predictability or whether they are simply due to internal variability of the atmosphere. By averaging over a large enough ensemble, any internal noise in each forecast should be averaged out, leaving a signal which is common across the ensemble and which, hopefully, represents genuine predictability. Of course, generating a large enough ensemble adds to the expense of running seasonal forecasts even more.

In order to run an ensemble of forecasts, perturbations to the initial conditions need to be generated. Many operational forecast systems overcome this problem by running their seasonal forecast ensemble once a month and using the initial analyses from the short range NWP forecast system for each day of the month (or a selection of days if less than 30 members are used) as the initial conditions for the ensemble members. The date of each analysis is reset to the first of the month and then these initial states are coupled to the ocean model and run as forecasts. In some seasonal forecasting systems the ocean initial conditions in the ensemble members are also perturbed, sometimes by adding random perturbations of the same magnitude as the uncertainty in specifying the ocean state and sometimes by spinning up the ocean model prior to coupling using perturbations to the surface winds and/or surface fluxes being used as the surface boundary condition.

7.4   Presentation of longer range forecasts

Forecasts beyond the predictability limit are usually expressed in deviations from normal conditions, over a reasonably long period. For instance, monthly forecasts may be presented in 10-day segments, or a seasonal forecast as a set of monthly anomalies, or just the anomaly over the whole season. The ensemble aspect of seasonal forecasting also allows the use of probabilities, so often forecasts may be presented in terms of probability of a particular variable being above or below normal, or in a particular quartile or decile. In contrast to short-range NWP forecasts there may be large areas of a forecast map which are blank as there is no discernible signal across the ensemble for that particular variable. Figure 7.6 shows a temperature prediction from the ECMWF seasonal forecast system. The forecast is for average 2 m temperature in the DJF(December, January, February) season of 2002/2003 from a forecast ensemble initiated in October 2002. The colours express the probability of the temperature being greater than the median value of temperature for that season. The winter of 2002/2003 saw a weak El Niño event in the equatorial Pacific and this is reflected in high probability of above-median temperatures across much of the tropics, not just the tropical east Pacific. There is also a strong signal for warmer than median temperatures in the southern hemisphere high latitudes. The typical North American pattern of warmer temperatures on the Pacific coast and cooler temperatures in the south-east United States can also be seen. Note that most of Europe and extra-tropical Asia has around a 50% probability of above-median temperatures (i.e. no real signal of either warmer or colder than normal conditions).

Figure 7.7 shows a similar forecast for precipitation in the same season as Figure 7.6. In this case the colours express the size of the forecast precipitation anomaly, in millimetres of precipitation during the season. It is notable that the area over which there is no discernible signal from the ensemble is much larger than that for the temperature forecast, implying that it is more difficult to get a significant signal of precipitation anomaly than it is for temperature.

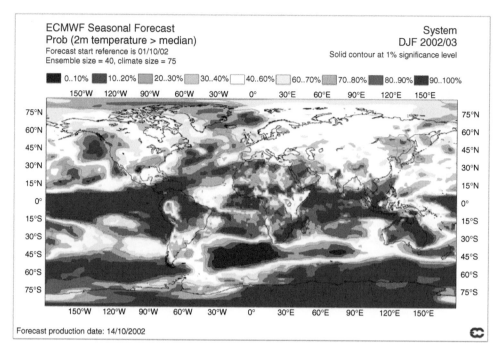

Figure 7.6 Two-metre temperature forecast from the ECMWF seasonal forecast system. The forecast is for the DJF season of 2002/2003, with colours expressing the probability of temperatures above the median value. (Reproduced by permission of ECMWF.)

In this case most of the significant signal is restricted to the tropical Pacific, with weaker signals across the rest of the tropics and very few strong signals outside the tropics.

Another common way of presenting seasonal forecast information is to give a most likely value and a range. This method can be applied to forecasts of variables such as mean monthly/seasonal temperature or precipitation. Several of the organisations which predict tropical storm activity also use this type of method. Figure 7.8 shows a forecast of Atlantic tropical storm activity from the UK Met Office. The forecast range is compared with the long-term average number of storms for this basin to give users some idea of whether the forecast activity is above or below normal.

As ENSO has such a big potential impact on seasonal forecasts, many forecast centres present forecasts of ENSO strength in some form. Simple graphs showing the evolution of the SST anomaly in the central or east Pacific (e.g. the NINO3 region, 5°S–5°N, 150°W–90°W) are an effective way of doing this, especially if an ensemble of forecasts or a group of forecasts from different models are available. Figure 7.9 shows such a forecast from

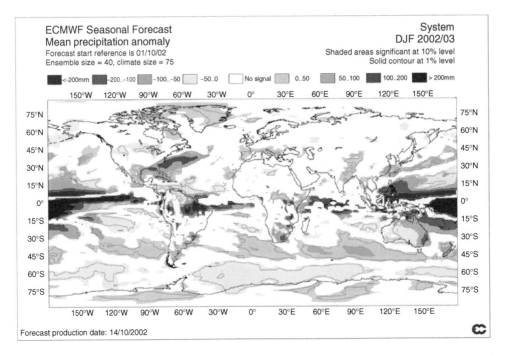

Figure 7.7 Precipitation anomaly forecast from the ECMWF forecast system for the DJF season of 2002/2003, initiated in October 2002. Colours represent the size of the forecast anomaly in millimetres over the season, with blue colours indicating wetter than normal and orange to red colours indicating drier than normal. (Reproduced by permission of ECMWF.)

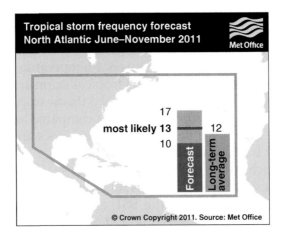

Figure 7.8   UK Met Office forecast of tropical storm numbers in the Atlantic for the 2011 storm season. (© Crown Copyright 2011, Met Office.)

NWS/NCEP/CPC

Last update: Tue Oct 25 2011
Initial conditions: 15Oct2011–18Oct2011

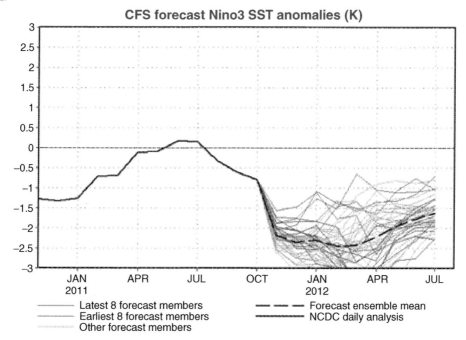

Figure 7.9 Forecasts of SST in the NINO3 region from the US NCEP seasonal coupled forecast system (CFS). Four forecasts are initiated every day and the figure shows the 40 most recent forecasts (i.e. those initiated in the past 10 days). The most recent forecasts are shown in blue. (Source: US National Centers for Environmental Prediction (NCEP).
http://www.cpc.ncep.noaa.gov/products/people/wwang/cfs_fcst/.)

the US NCEP seasonal forecasting system. The thick black line shows the observed SST in the NINO3 region through 2011 and the thinner coloured lines show the forecast from each ensemble member initiated in October 2011. The thick black dashed line is the ensemble mean. All the ensemble members indicate a rapid drop in NINO3 SST during the early boreal winter of 2011 – that is the onset of La Niña. The most recent forecasts are shown in blue and these all indicate warmer temperatures than the earlier forecasts (shown in red), perhaps indicating that the predicted cooling may not be as strong as predicted by the ensemble mean.

Because of the large uncertainty and lower levels of predictable signal in seasonal forecasts, most major forecasting centres which produce seasonal forecasts tend to rely on output from several different models, some empirical others using NWP methods. Some centres, such as the International Research Institute for Climate and Society (IRI) at Columbia University, compile summaries of the seasonal predictions from all the major forecast centres. These are accompanied by a discussion of which models have been used in the predictions and how those forecasts have been weighted to produce the summary (http://iri.columbia.edu/climate/forecast/net_asmt/).

Summary

- Despite the deterministic limit on weather forecasting there are still many customers who require forecast information on lead times of greater than 14 days. Forecasting beyond the deterministic limit is also an interesting scientific challenge.
- There are several sources of predictability for forecasts on periods of months to seasons. These sources include sea surface temperature anomalies (especially in the Tropics), subsurface ocean temperature anomalies, land surface snow, ice and soil moisture anomalies, stratospheric conditions and solar variability.
- Monthly to seasonal forecasts can either be *empirical* – based on statistical relationships between predictor variables and a particular quantity being forecast – or *numerical* – using NWP techniques.
- Empirical forecast methods can be simple and cheap to build and use. However, they can be subject to a number of different problems, including the tendency of statistical relationships between meteorological variables to be non-stationary in time. There may also be significant seasonal variations in weather which are not related to any of the precursor variables used in the model.
- Numerical methods are much more expensive, usually requiring the coupling of an atmospheric NWP model to some form of ocean model and the use of ensemble methods.
- The bias of an NWP seasonal forecast tends to increase with forecast lead time. This needs to be dealt with through the generation of forecast lead-time dependent model climatologies. These are expensive to produce, requiring many years of model hindcasts to be run.
- The presentation of monthly to seasonal range forecasts tends to be done in terms of the probability of departure of the forecast variables from a long-term climatological mean.

8

Forecast Verification

It is natural that we should want to regularly assess performance in forecasting the weather. There are many reasons for doing so, of benefit to both
providers and users of forecasts. For example, *providers* need to monitor
trends in forecasting performance, measure the impact of model upgrades, be
able to recognise those circumstances when a model is more or less reliable
and accordingly prioritise further research and development. Armed with
knowledge of forecast performance, *users* could meanwhile use the information to select between providers and assess whether forecast accuracy meets
their own requirements in terms of value. Both *providers* and *users* might
reasonably ask the question 'Are the forecasts skilful?' Providers and users
might differ in their natural initial attention, respectively, to spatial average
and site-specific performance. Development of new approaches to forecast
verification remains an active area of research and discussion, including dedicated research programmes (e.g. the World Weather Research Programme,
WWRP), conferences and/or symposia (e.g. the series of International Conferences on Statistical Climatology and of Verification Methods Workshops)
and special issues of journals (e.g. *Meteorological Applications*). Nor, of course,
is forecast verification an activity restricted to weather forecasting and climate prediction, with much cross-over activity in other disciplines where
forecasting is also important.

Before we go any further, it is necessary to define exactly what we mean
by some terms we have already used, colloquially, in the above introduction:
accuracy, *value* and *skill*:

Forecast accuracy (or quality) – a qualitative or quantitative measure of the
extent to which a forecast simulates observed conditions.

Forecast value – the extent to which a forecast helps a user to take a better
decision.

Operational Weather Forecasting, First Edition. Peter Inness and Steve Dorling.
© 2013 John Wiley & Sons, Ltd. Published 2013 by John Wiley & Sons, Ltd.

Forecast skill – a quantification of the improvement in forecast accuracy relative to a simpler baseline method requiring no particular expertise (for example *Persistence* or *Climatology* – Box 8.1).

Box 8.1 Persistence and Climatology

Persistence – The simplest type of forecast is one which assumes 'more of the same', that is a repeat tomorrow of the conditions most recently experienced. The success of such an approach will depend upon how far into the future we are forecasting and, indeed, upon the location – in general, a medium-range 5–10 day persistence forecast in a changeable, mid-latitude climate such as Ireland would be much less likely to be accurate than a one-day persistence forecast in the middle of the winter monsoon in China. The success of a persistence forecasting approach is also 'flow dependent', even in changeable climates, such that there will be periods when the large scale circulation exhibits consistent behaviour over days, weeks or even, at the extreme, months – we will see later what impact this can have on the statistical performance of NWP forecasts. In general, though, persistence forecasting requires no special knowledge and our NWP model needs to outperform this approach when evaluated over a period if we are to demonstrate its usefulness.

Climatology – We all, almost without knowing, apply the concept of climatology when we pack a suitcase to go on holiday. If we are travelling to a fairly familiar place, we already have an idea what sort of weather is common, or possible, from previous visits, consideration of the time of year and perhaps from what we have read or heard about the place over the years. Choice of clothing will, at least in part, be a response to our awareness of the typical climatology of the place; indeed, climate is sometimes described as the conditions you expect to experience. If we are visiting a less familiar destination, we might consult a travel guide or the climate normal for a particular location, such as the one shown in Figure B8.1.1 for Edinburgh in Scotland. Climate normals are based on thirty years of station records, in this case 1971–2000, the idea being that such a length of record captures, over the most recent decades, the natural variability which occurs both through the year and between years. The climate normal for a location is a good baseline source of a 'first guess' of the weather conditions that might be expected. If our NWP forecast cannot improve on climatology, when assessed over a reasonable period, then we would say that the NWP approach does not exhibit skill and we would question its purpose, unless the NWP forecast added value for a specific set of circumstances of importance to a user.

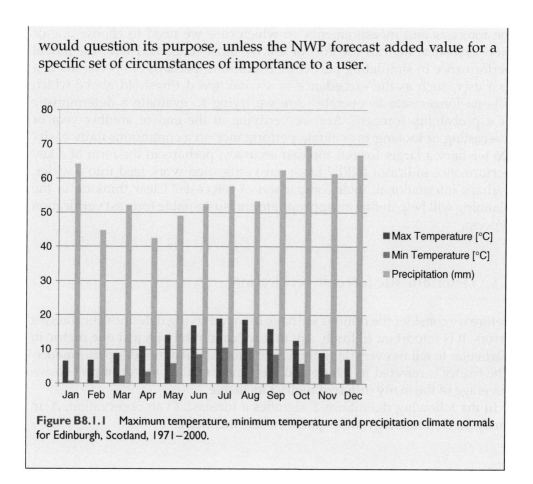

Figure B8.1.1   Maximum temperature, minimum temperature and precipitation climate normals for Edinburgh, Scotland, 1971–2000.

Even with these definitions in place, there are still many issues to carefully consider and settle on before we rush to evaluate model performance. We must acknowledge that it is not viable to monitor, comprehensively, every component of model output. For one thing, as the earlier chapter on observations showed, we do not have observational 'truth' everywhere against which to compare our forecasts, but the need for long-term undisturbed observational records to support forecast verification procedures is clear. Are we going to verify the raw gridded model output or some form of post-processed forecast which already takes advantage of observational data through some form of downscaling approach? Will we focus on the surface variables most familiar to users, such as the daily surface maximum temperature at a particular location, or consider the accuracy of the forecast in capturing the 500 hPa circulation over a particular geographical domain? Are we most concerned with regularly quantifying the differences between the forecasts and measurements, in which case we need to choose one or more statistical metrics for this, or are we most concerned with the model performance in simulating particular conditions, perhaps of special interest to a user, such as the exceedance of a wind speed threshold above which it is no longer safe to operate? Are we trying to evaluate a deterministic or a probability forecast? Are we verifying at the end of another year of forecasting or looking to evaluate performance on a continuous daily basis? Do we have a target for our forecast accuracy, perhaps in the form of a key performance indicator (KPI)? Does our verification work feed into a wider, perhaps international, inter-comparison of forecasts? Clear thinking in the planning will help define an appropriate and sustainable forecast verification process.

8.1   Deterministic forecast verification

Before we consider the options we have for quantifying deterministic forecast errors, it is important to firstly say that we cannot rely on just one metric in particular to tell us everything there is to learn about a model's performance. The reader is referred to Jolliffe and Stephenson (2012) for a comprehensive coverage of the many different metrics available.

In the following definitions F signifies a forecast, O an observation, A an analysis, C climatology and $^-$ (an over-bar) the mean.

Bias

$$Bias = \sum_{i=1}^{n} \overline{F_i - O_i}$$

Range $-\infty$ to ∞; Perfect Score $= 0$

Bias is the measure of forecast performance with which people would tend to feel most familiar. It quantifies the tendency for forecasts to over- or underestimate observations. The nature of any model bias can be effectively revealed in a scatter plot of forecasts against observations in which it may be shown that any bias may, in fact, not be systematic but flow dependent or vary with, for example, season or time of day. If model biases are well understood then corrections can be straightforwardly applied to produce a more accurate site-specific forecast. The cancellation of model biases of different sign can result in a misleading assessment of forecast performance when a more superficial approach is adopted, so it is recommended that bias assessment is combined with other metrics to avoid this scenario.

Mean Absolute Error (MAE)

$$MAE = \sum_{i=1}^{n} \overline{|F_i - O_i|}$$

Range 0 to ∞; Perfect Score = 0

Cancellation of errors of different sign is avoided by considering the MAE. Errors with different magnitude are given equal weighting.

Root Mean Square Error (RMSE)

$$RMSE = \sum_{i=1}^{n} \sqrt{(F_i - O_i)^2}$$

Range 0 to ∞; Perfect Score = 0

The RMSE, based as it is on the square of the model error, penalises large errors more heavily and, at a practical level, this may be argued to be a more appropriate way to weight errors compared with the MAE approach.

Figure 8.1 is an inter-comparison between global models from different forecast centres showing RMSE of surface pressure in the Northern Hemisphere

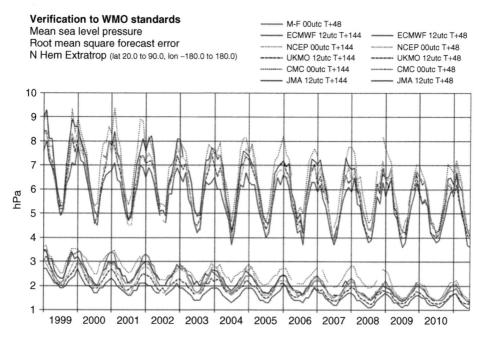

Verification to WMO standards
Mean sea level pressure
Root mean square forecast error
N Hem Extratrop (lat 20.0 to 90.0, lon −180.0 to 180.0)

M-F 00utc T+48
ECMWF 12utc T+144   ECMWF 12utc T+48
NCEP 00utc T+144   NCEP 00utc T+48
UKMO 12utc T+144   UKMO 12utc T+48
CMC 00utc T+144   CMC 00utc T+48
JMA 12utc T+144   JMA 12utc T+48

hPa

Figure 8.1 Global Model RMSE Surface Pressure for Northern Hemisphere Extra-tropics based on T+48 and T+144 forecasts, 1999–2011. (Reproduced by permission of ECMWF.)

Extra-tropics, for T+48 and T+144 time horizons, neatly demonstrating both seasonal dependence and long-term improvements. Each model is verified against its own analysis.

Anomaly Correlation Coefficient (ACC)

$$ACC = \frac{\overline{(F-C)(A-C)}}{\overline{(F-C)^2(A-C)^2}}$$

Range -100 to 100%; Perfect Score $= 100\%$

Simple spatial correlation of forecasts with observations or analyses may give an impression of unrealistic accuracy due to natural seasonal variations which inflate the model performance. Climate averages are, therefore, subtracted from both forecasts and verifications and the resulting anomalies are then verified in the form of an anomaly correlation coefficient. As such the ACC is a positively orientated *skill score*, meaning that high values are good and that this metric already accounts for climatology and, therefore, quantifies added value.

An ACC of 50% is equivalent to a baseline climatology performance, while 60% and 80% represent skilful model performances at the largest weather pattern scales and at the synoptic scales respectively. Figure 8.2 shows the forecast range at which a 500 hPa geopotential height ACC of 80% has been achieved using the ECMWF Deterministic Model for a European domain,

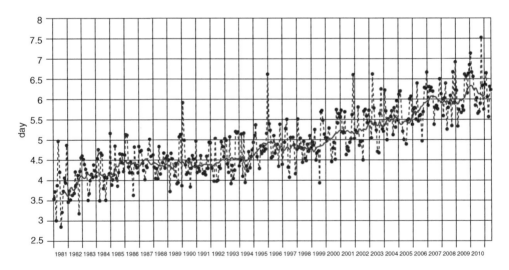

Figure 8.2   Forecast time horizon at which ACC of the 500 hPa geopotential height forecast first falls below 80% based on the ECMWF Deterministic Model for a Europe domain over the period 1981–2011. Blue dashed line shows monthly mean performance, the red solid line a twelve-month running mean. (Reproduced by permission of ECMWF.)

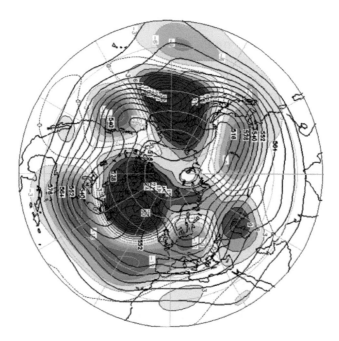

Figure 8.3   Mean monthly Northern Hemisphere 500 hPa Geopotential Height (black contours) and anomaly (colour scale) for December 2010, based on CDAS Reanalysis data. (Image courtesy of the NOAA Climate Prediction Center; http://www.cpc.ncep.noaa.gov/.)

from 1981 to 2011, revealing both long-term model improvements and month-to-month flow dependency. The highest monthly mean performance, in December 2010, was associated with a persistent blocked upper air pattern in the Atlantic, the monthly mean 500 hPa geopotential anomaly pattern for which is shown in Figure 8.3.

Skill scores

As already stated, the skill of a forecast is measured by comparing its performance with that of some basic level of forecasting which could be assumed to have zero skill, such as persistence (forecasting the same weather as yesterday) or climatology (forecasting the climatological normal conditions). A *skill score* is computed by comparing a forecast statistic, such as RMSE or ACC, for the forecasting system under test with the same measure for the zero skill method in the following way:

$$skill\ score = 1 - \left[\frac{RMSE_f}{RMSE_{ZS}} \right]$$

where the subscript f indicates the value for the forecasting system and the subscript ZS indicates the value for the zero skill method. This will take a value of one for perfect forecasts and zero for forecasts which are no more or less skilful than the zero skill method.

In situations where the weather remains very similar for many days or even weeks, skill scores based on persistence will suffer, with models apparently having reduced skill simply because persistence is actually providing rather good forecasts. This is an important consideration to bear in mind when designing metrics for assessing forecast performance, particularly if there is a financial penalty or cost involved for forecasts which appear to perform badly.

Contingency tables

When we want to evaluate the performance of a forecast model in predicting the occurrence of a specific, pre-defined event, such as a ground frost, a tornado or rainfall exceeding a certain threshold, we need an alternative approach. In these circumstances, there are four possible outcomes, as show in the contingency table (Table 8.1).

From the contingency table, a number of performance measures can be defined.

$$Frequency\ Bias,\quad FBI = \frac{(A+B)}{(A+C)}$$

Frequency Bias quantifies whether the forecast system tends to predict the particular event in question more or less often than it is actually observed, with values less than one indicating that the forecast system tends to under-predict the occurrence and values greater than one indicating a tendency to over-predict occurrence.

$$Hit\ Rate\ (Probability\ of\ Detection),\quad HR\ (POD) = \frac{A}{(A+C)}$$

Table 8.1   Structure of a typical contingency table

Event forecasted?	Event observed?		
	YES	NO	Marginal total
YES	HIT	FALSE ALARM	Forecasted
NO	MISS	CORRECT NULL	Not forecasted
Marginal total	Observed	Not Observed	Sum Total

Event forecasted?	Event observed?		
	YES	NO	Marginal total
YES	A	B	$A+B$
NO	C	D	$C+D$
Marginal total	$A+C$	$B+D$	$A+B+C+D=n$

The *Hit Rate* quantifies the fractional success of the forecast system in predicting the particular event on the occasions when it actually occurs, with a score of one indicating a perfect Hit Rate. Clearly, a forecast system which successfully forecasts most or all events of interest is very desirable, particularly if those events are likely to have a large impact. However, this can easily be achieved by *always* predicting such an event, or at least significantly over-predicting its occurrence. Whilst achieving a high Hit Rate, this type of forecast would have little value to a customer. In fact, over-confident predictions of this type could be very financially detrimental if the customer has to take expensive action in response to a positive forecast of a particular event, such as closing down production on an oil rig or applying salt to roads.

As well as having a measure which quantifies the ability of a forecast system to predict a particular event, it makes sense, therefore, to also have a measure of how many times that particular event is forecast on occasions when it *doesn't* actually occur – that is the false alarms:

$$\text{False Alarm Rate, } FAR = \frac{B}{(B+D)}$$

Of course, it is possible to achieve a perfect *False Alarm Rate* of zero simply by never forecasting the particular event, but that would also lead to a Hit Rate of zero. To a customer, what matters most is the economic impact of the forecasts – do they help the user to either save money or make profit? This will depend on a number of factors:

- What is the cost of taking action in response to a particular forecast? For example, how much does it cost to spread salt on the roads if ice is forecast?
- What is the cost (or loss) incurred by not taking action if the event actually occurs? For example, what is the cost of damage to drilling equipment on an oil rig if it is not secured prior to a storm?
- How much extra profit can be generated by taking action in response to a correct forecast? For example, how much money would an ice-cream salesman make by taking an ice-cream van to the beach on the strength of a forecast of warm, sunny weather?

These factors will differ for every customer, so different customers would be prepared to accept different values of Hit Rate and False Alarm Rate.

The *Relative Operating Characteristic* (ROC) score is a concept taken from signal processing. It combines Hit Rate (HR) and False Alarm Rate (FAR) by plotting them on an x–y diagram with FAR on the x-axis and HR on the y-axis. A set of perfect forecasts would appear as a point in the top left corner of the diagram – that is FAR = 0 and HR = 1. Completely random forecasts would appear as a point right in the centre of the diagram (HR = FAR = 0.5). Forecasts with a degree of skill (i.e. better than random) would appear as points above the diagonal from bottom left to top right, and forecasts with negative skill (i.e. worse than random) would appear as a point below this diagonal.

Equitable scores are particularly valuable in the verification of deterministic forecasts, since they heavily penalise constant and purely random forecasts (Gandin and Murphy, 1992). Equitability requires that all random forecasting systems should attract the same score. The often seen *Equitable Threat Score*, otherwise known as the *Gilbert Skill Score*, does not in fact adhere to this definition:

$$\text{Equitable Threat Score, } ETS = \frac{(A - A_R)}{(A + B + C - A_R)}$$

where

$$A_R = \frac{(A + B)(A + C)}{n}$$

The commonly seen *Peirce's Skill Score*, on the other hand, is a truly equitable measure:

$$\text{Peirce's Skill Score, } PSS = \frac{(AD - BC)}{(A + C)(B + D)} = POD - FAR$$

$$1 - PSS = \text{Miss rate} + FAR$$

Despite the nineteenth century origin of the contingency table, research is still continuing to identify new skill scores for particular applications. For example, although PSS is truly equitable, it is not helpful when evaluating some precipitation forecasts because it is restricted to two categories when, in practice, we might wish to separate dry forecasts from a number of other amount categories. The equitable SEEPS score (Stable Equitable Error in Probability Space) was introduced by Rodwell *et al.* (2010) to address this issue, using 'dry', 'light precipitation' and 'heavy precipitation' as categories, where the light and heavy categories are defined on a site-specific basis using station climatology and the light/heavy boundary is determined by the rule that light is twice as frequent as heavy. SEEPS is written as the mean of two two-category scores, which respectively relate to the dry/light and light/heavy boundaries. Each of these scores can be written as:

$$(\text{Misses/Expected events}) + (\text{False Alarms/Expected non-events})$$

So a perfect SEEPS score, with no misses or false alarms, is zero. Figure 8.4 shows a case study application of SEEPS, showing its usefulness in highlighting geographical areas of relatively high SEEPS values and, thereby,

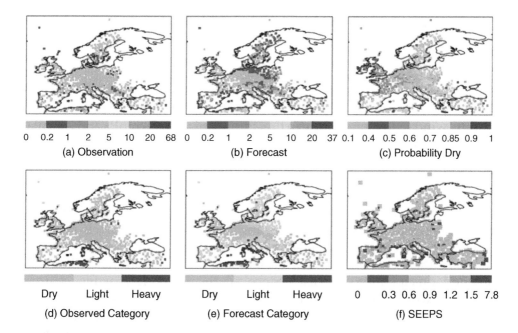

0   0.2   1   2   5   10   20   68   0   0.2   1   2   5   10   20   37   0.1   0.4   0.5   0.6   0.7   0.85   0.9   1
(a) Observation   (b) Forecast   (c) Probability Dry

Dry   Light   Heavy   Dry   Light   Heavy   0   0.3   0.6   0.9   1.2   1.5   7.8
(d) Observed Category   (e) Forecast Category   (f) SEEPS

Figure 8.4   (a) Observed precipitation (mm) accumulated over the 24 hours to 12:00 UTC on 23 April 2010. (b) Forecast precipitation (mm) accumulated over lead-times 0–24 hours and valid for the same period as the observations. (c) Probability of a 'dry' day in April based on the 1980–2009 climatology. (d) Observed precipitation category. (e) Forecast precipitation category. (f) SEEPS. (Reproduced from Rodwell, MJ *et al.* *QJR Meteorol Soc* 2010;136:1344–1363, with permission from John Wiley & Sons, Ltd.)

drawing attention to synoptic features or processes which are poorly handled by the model.

8.1.1   WMO verification metrics

The World Meteorological Organization (WMO) publishes regular verification statistics for all Weather Forecasting Centres which run global NWP model. Figure 8.1 is an example of the type of verification measures published by WMO. These measures are more than just a beauty contest between the major forecasting centres. They help forecasting organisations and individual forecasters to make informed decisions about which numerical models perform best in particular regions and for particular variables. They also allow forecasting centres to measure the performance of their own model against those of other centres and potentially share information on how improvements to model performance have been achieved.

The categories that the WMO uses for its verification metrics are:

- **Region:** Northern hemisphere extra-tropics (20°N–90°N), Southern hemisphere extra-tropics (20°S–90°S) and the tropics (20°N–20°S).
- **Variable:** pressure, geopotential height, temperature and winds.
- **Level:** mean sea level for pressure; 500 hPa and 250 hPa for geopotential height, winds and temperature in the extra-tropics; and 850 hPa and 250 hPa for geopotential height, winds and temperature in the tropics.
- **Forecast lead time:** T+24 and every 24 hours thereafter.

These metrics give a very broad indication of model performance over wide regions. Individual forecasting centres will almost certainly compile their own more detailed verification statistics for how their models perform in their particular area of interest and for variables which may be of more interest to their customers, such as temperature forecasts for individual cities and precipitation forecasts. Since all the major forecasting centres around the world are funded from central government in some way, most centres are required to submit these verification measures to government scrutiny. In some cases, performance targets will be set based on these verification measures and the performance-related pay of individuals within the forecasting organisation may even be dependent on meeting these targets.

8.2   Verification of probability forecasts

The verification of probability forecasts calls for sophisticated techniques to determine the usefulness or skill of the forecasts. Naively one might assume that a probability forecast can never be wrong so long as the 0% and 100% probability categories are avoided. If a forecast is issued for a 10% probability of rain and rain actually occurs, is that forecast right or wrong? A forecast customer who decided to hold a particular outdoor event on the basis of that forecast might feel somewhat aggrieved when rain actually occurred, but the forecast did allow for the (albeit small) probability that rain would occur.

The only way to perform any kind of objective verification of probability forecasts is to look at their performance over an extended period. Consider the example above of a forecast of a 10% probability of rain. Over a long period, the only reasonable expectation is that rain should occur on 10% of the occasions when it was forecast with that level of probability. If it actually rains on more than 10% of the occasions when rain was forecast with a 10% probability, then the forecasts can be said to be *over-confident*, being too sure that rain is not very likely. However, if rain occurs on less than 10% of those occasions then the forecasts can be said to be *under-confident*, being reluctant to forecast a lower or zero probability.

8.2.1   Reliability diagrams and the Brier Score

For each probability bin used in the forecasts, the observed frequency of the particular event being forecast can be calculated from a long record of events. The observed frequency can be plotted against the forecast probability to give a graphical representation of the forecast performance. An example of such a reliability diagram is shown in Figure 8.5. The best possible set of forecasts would lie perfectly on the bottom-left to top-right diagonal of this diagram (the blue line), indicating that the forecasts realistically represented the observed frequency in all probability categories. The assumption is that both the ensemble members and measurements are sampled in the same way from the underlying probability distribution.

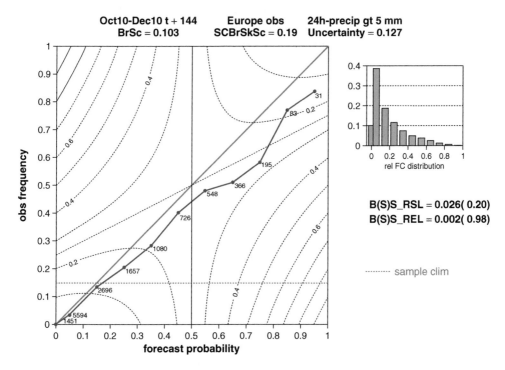

Figure 8.5   Reliability diagram for T+144 ECMWF ensemble prediction system forecasts of >5 mm 24-hour precipitation totals for European observing stations, October–December 2010. (Courtesy of Anna Ghelli. Reproduced by permission of ECMWF.)

The area between the forecast curve (the red line in Figure 8.5) and the 1:1 diagonal (the blue line in the figure) can be used as a quantitative measure of the *reliability* of probability forecasts, with a low value indicating reliable forecasts. The reliability can be written down in the form of an equation as:

$$Reliability = \frac{1}{N} \sum_{i=1}^{N} n_i(f_i - o_i)^2$$

where N is the total number of forecasts, n_i is the number of forecasts in each probability bin, f_i is the forecast probability and o_i is the observed frequency of the event when forecast with probability f_i. A perfectly reliable forecast would have a score of zero. In the example shown in Figure 8.5 the reliability is 0.002.

Another measure that can be inferred from probability forecast verification data is the *resolution* of the forecasts. This gives an indication of how good the forecasts are at predicting the occurrence of an event with a forecast probability which is very different from its climatological frequency. So, for an event which occurs infrequently (say 10% of the time) the resolution score will reward forecasts which predict the occurrence of that event with a very high probability. In terms of area on the reliability diagram, the resolution is the area between the forecast curve (the red line) and the horizontal climatological frequency value (the dotted blue line in Figure 8.5). In this example the climatological frequency of the event is quite low (about 0.15), so it is clear from the figure that the largest contribution to the resolution score comes from forecasts which predict the event with a much higher probability than the climatological value. The resolution can also be calculated using:

$$Resolution = \frac{1}{N} \sum_{i=1}^{N} n_i(o_i - c)^2$$

where c is the climatological frequency of the event being forecast and the other symbols have the same meaning as in the reliability calculation. A high resolution score indicates useful forecasts. For perfect forecasts of an event with a climatological frequency of 0.5 the resolution would be 0.25. For the example in Figure 8.5 the resolution is 0.026. Forecasts with zero resolution would be indicated by a horizontal line on the reliability diagram, with the observed frequency being equal to the climatological value for every single forecast probability bin. This would mean that the forecast method is unable to recognise when the event in question was more or less likely to occur.

Reliability and resolution can be combined, along with a measure of the *uncertainty* of the particular event being forecast, into the *Brier Score*. Uncertainty is simply given by:

$$Uncertainty = c(1-c)$$

So that events which have a climatological frequency of 0.5 have the highest possible uncertainty (0.25) and events that either occur all the time or never occur have an uncertainty of zero.

The Brier Score (BS) is given by:

$$BS = Reliability - Resolution + Uncertainty$$

A Brier Score of zero indicates perfect, deterministic forecasts and a score of one indicates perfectly wrong deterministic forecasts. If every single forecast that was issued gave a probability of 50% for the occurrence of the event in question the Brier Score would be 0.25. This means that a Brier Score of less than 0.25 indicates a degree of skill in the forecasts. The Brier Score for the example shown in Figure 8.5 is 0.103.

8.2.2   Relative Operating Characteristic (ROC) for probability forecasts

Forecast centres are increasingly using the ROC approach to quantify probability forecast performance. Hit Rate (HR) and False Alarm Rate (FAR) values can be generated for *each probability threshold* in the forecasts. The ROC score for each probability threshold can be plotted as a point on the ROC diagram, leading to an ROC *curve*. Figure 8.6 shows an example of how an ROC curve can be constructed from a contingency table of forecast and observed data. The (trapezoidal) area between the ROC curve and the 1:1 line is a measure of forecast *skill*, with a large area indicating high skill and a zero area indicating zero skill. The upper limit would be for perfect deterministic forecasts giving a ROC area of 0.5. Figure 8.7 shows an example, based on precipitation forecasts, of how the ROC area can change over time as a function of weather type, season and model improvements.

8.3   Subjective verification

The methods described above show how model performance for either deterministic or probability forecasts can be quantified objectively. The metrics used are useful for informing NWP model developers on areas of model performance that need to be targeted. Model developers also receive subjective feedback from both forecasters and customers on the performance of the NWP model in particular circumstances, and this type of information too is very useful. For instance, forecasters providing services to military aviation may have noticed that the NWP model always over- or under-predicts low

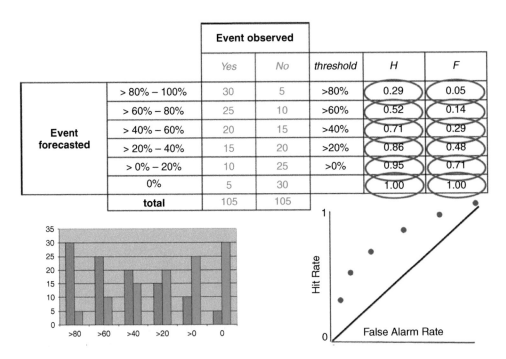

Figure 8.6 A schematic example of a ROC curve generated from contingency table results for probability forecasts. (Reproduced by permission of ECMWF.)

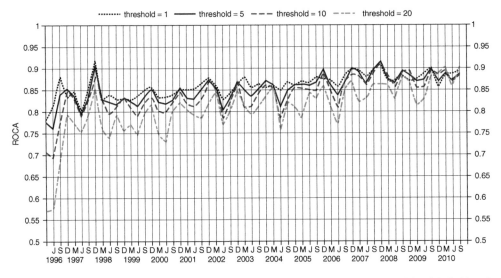

Figure 8.7   ROC Area T+96 forecasts of 24-hour precipitation totals for thresholds of 1, 5, 10 and 20 mm based on ECMWF forecasts

cloud amounts in a particular area of operations. This type of subjective verification by forecasters who are using the NWP output helps model developers prioritise issues which need to be improved. Frequent comments by a large number of forecasters or customers indicate issues which are clearly important to the users of the NWP output, so need to be dealt with quickly. Very specific issues, such as the example of low cloud amounts, also give model developers clues as to which aspects of the model need to be worked on to fix the problem. Whereas a systematic bias in sea level pressure forecasts is difficult to relate to any particular component of the numerical model, an error in low cloud amounts may be related to problems with the cloud scheme or the temperature structure of the boundary layer in the model.

The *value* of a probability forecast will vary depending on the user of that forecast and how financially sensitive their operations are to particular weather conditions. Box 8.2 discusses the issue of forecast value in more detail.

Box 8.2 Value of probability forecasts

The value of a weather forecast can be described as the extent to which a forecast helps a user make a better decision. Forecasts can also be assessed in terms of their economic value and whilst probability forecasts might seem initially to make it harder to make decisions, they really become useful when considering their economic potential. The best way to show this is to consider a very simple example.

A restaurant has space for outside tables in front of the premises but these can only be used if the weather is dry and warm (temperature greater than 20°C).

If the outside tables are in use, the restaurant will generate $600 extra income per day, but the owner will need to employ an extra member of staff at a cost of $60 per day.

If a probability forecast of the necessary conditions for using the outside tables is available (i.e. the probability that there will be no rain and temperature greater than 20°C), at what forecast probability level ought the restaurant owner to consider hiring the extra member of staff?

A naïve answer would be greater than 50% (i.e. there is a better than even chance of good weather). However, as long as the forecasts are unbiased the owner would actually make a profit over a long period if they hired the extra member of staff on the basis of forecasts with a probability of greater than 10%. This is because the extra profit to be made on the occasions when the weather is good is 10 times the cost of hiring the extra staff member. If the extra costs involved were higher, then the restaurant owner should take action at a higher probability. If, for instance, the extra staff cost $120 but the profit remained the same, then the owner should hire the extra member of staff at probabilities greater than 20%. For a business with a much higher ratio of cost to profit, the probability at which it ought to take action would also be much higher, with the probability at which to take action being very simply given by the cost/profit ratio.

Summary

- The verification of forecasts is an important aspect of the work of a forecasting centre, feeding into the development and improvement of forecast models.
- Verification is also important to customers who can make informed choices based on the skill of different forecast providers, and can also modify their weather-dependent behaviour based on the known characteristics of the forecasts they are using.
- The World Meteorological Organization (WMO) oversees a verification of the global NWP models from all the major forecasting centres in the world, using a range of different domains, forecast variables and lead times.
- Forecast skill measures the ability of a particular forecasting method to improve upon forecasts made using some *zero-skill* technique such as persistence or climatological forecasts.
- For simple deterministic forecasts predicting the occurrence or non-occurrence of a particular meteorological condition, contingency tables can generate a wide range of verification scores which can be of particular relevance to forecast customers.
- Verification of probability forecasts is complex, but the use of *reliability diagrams* and the *Brier Score* can provide useful quantification of forecast performance.
- Probability forecasts can help customers to make economic decisions on the basis of the weather forecast.

References

Arakawa, A. and Schubert, W.H. (1974) Interaction of a cumulus cloud ensemble with the large-scale environment, Part I. *J. Atmos. Sci.*, **31**, 674–701.

Baldwin, M.P. and Dunkerton, T.J. (2001) Stratospheric harbingers of anomalous weather regimes. *Science*, **294**, 581–584.

Balling Jr, R.C., Vose, R.S. and Weber, G. (1998) Analysis of long-term European temperature records: 1751–1995. *Climate Research*, **10**, 193–200.

Berthelot, M., Dubus, L. and Gailhard, J. (2011) Improvement of Meteorological and Hydrological Probabilistic Monthly Forecasts over France with an Analogue Method. International Conference Energy and Meteorology (ICEM), 8–11 November 2011, Gold Coast, Australia.

Bjerknes, J. (1966) A possible response of the atmospheric Hadley circulation to equatorial anomalies of ocean temperature. *Tellus*, **18**, 820–829.

Bjerknes, J. (1969) Atmospheric teleconnections from the equatorial Pacific. *Mon. Wea. Rev.*, **97**, 163–172.

Branski, F. (2010) Pioneering the collection and exchange of meteorological data. *WMO Bulletin*, **59**, 12–17.

Cardinali, C., Pezzulli, S. and Andersson, E. (2004) Influence-matrix diagnostic of a data assimilation system. *Q.J.R. Meteorol. Soc.*, **130**, 2767–2786.

Carlson, T.N. (1980) Airflow through mid-latitude cyclones and the comma cloud pattern. *Mon. Wea. Rev.*, **108**, 1498–1509.

Carroll, E.B. (1997) A technique for consistent alteration of NWP output fields. *Meteorol. Apps.*, **4**, 171–178.

Carroll, E.B. and Hewson, T.D. (2005) NWP grid editing at the Met Office. *Weather and Forecasting*, **20**, 1021–1033.

Charney, J.G., Fjørtoft, R. and von Neumann, J. (1950) Numerical integration of the barotropic vorticity equation. *Tellus*, **2**, 237–254.

Collard, A., Hilton, F., Forsythe M. and Candy, B. (2011) From Observations to Forecasts – Part 8: The use of satellite observations in numerical weather prediction. *Weather*, **66**, 31–36.

Donlon, C. J., Martin, M., Stark, J.D., Roberts-Jones, J., Fiedler, E. and Wimmer, W. (2011) The Operational Sea Surface Temperature and Sea Ice Analysis (OSTIA). *Remote Sensing of the Environment*. doi: 10.1016/j.rse.2010.10.017.

Operational Weather Forecasting, First Edition. Peter Inness and Steve Dorling.
© 2013 John Wiley & Sons, Ltd. Published 2013 by John Wiley & Sons, Ltd.

Eden, P. and Burt, S. (2010) Extreme rainfall in Cumbria, 18–20 November (2009). *Weather*, **65**, 14.

Ferranti, L. and Corti, S. (2011) New clustering products. *ECMWF Newsletter*, **127**, 6–11.

Gandin, L.S. and Murphy, A.H. (1992) Equitable skill scores for categorical forecasts. *Mon. Wea.Rev.*, **120**, 361–370.

Hoskins, B. J. and Simmons, A.J. (1975) A multi-layer spectral model and the semi-implicit method. *Q.J.R. Meteorol. Soc.*, **101**, 637–655.

Jolliffe, I.T. and Stephenson, D.B. (eds) (2012) *Forecast Verification – A Practitioner's Guide in Atmospheric Science*, 2nd edn. John Wiley & Sons Ltd, Chichester.

Kalnay, E. (2003) *Atmospheric modeling, data assimilation and predictability*. Cambridge University Press.

Kelly, G. and Thépaut, J-N. (2007a) Evaluation of the impact of the space component of the Global Observing System through Observing System Experiments. *Seminar on recent developments in the use of satellite observations in numerical weather prediction, 3–7 September 2007*. Available at: http://www.ecmwf.int/publications/library/do/references/show?id=88345.

Kelly, G. and Thépaut, J-N. (2007b) The relative contribution of the various space observing systems. In: *Joint Eumetsat/AMS Conference Proceedings*, 24–28 September 2007, Amsterdam, The Netherlands (http://www.eumetsat.int/home/Main/AboutEUMETSAT/Publications/ConferenceandWorkshopProceedings/2007/groups/cps/documents/document/pdf_conf_p50_s2_01_kelly_v.pdf; accessed 20 June 2012).

Klotzbach, P. J. (2008) Refinements to Atlantic basin seasonal hurricane prediction from 1 December. *J. Geophys. Res.*, **113**, D17109.

Leutbecher, M., Barkmeijer, J., Palmer, T.N. and Thorpe, A.J. (2002) Potential improvement to forecasts of two severe storms using targeted observations. *Q.J.R. Meteorol. Soc.*, **128**, 1641–1670.

Lockwood, M., Harrison, R.G., Woollings, T. and Solanki, S.K. (2010) Are cold winters in Europe associated with low solar activity? *Env.Res. Lett.*, **5**, 024001.

MacDonald, A.E., Xie, Y. and Ware, R.H. (2002) Diagnosis of Three-Dimensional Water Vapor Using a GPS Network. *Mon. Wea. Rev.*, **130**, 386–397.

Manabe, S., Smagorinsky, J. and Strickler, R. F. (1965) Simulated climatology of a general circulation model with a hydrologic cycle. *Mon. Wea. Rev.*, **93**, 769–798.

Manley, G. (1974) Central England Temperatures: monthly means 1659 to 1973. *Q.J.R. Meteorol. Soc.*, **100**, 389–405.

Manton, M.J. and Warren, L. (2011) A Confirmatory Snowfall Enhancement Project in the Snowy Mountains of Australia. Part II: Primary and Associated Analyses. *J. Appl. Meteor. Climatol.*, **50**, 1448–1458.

Manton, M.J., Warren, L., Kenyon, S.L., Peace, A.D., Bilish, S.P. and Kemsley, K. (2011) A Confirmatory Snowfall Enhancement Project in the Snowy Mountains of

Australia. Part I: Project Design and Response Variables. *J. Appl. Meteor. Climatol.*, **50**, 1432–1447.

Meischner, P. (2010) *Weather Radar: Principles and Advanced Applications*. Springer.

Mohr, T. (2010) The Global Satellite Observing System: A Success Story. *WMO Bulletin*, **59**, 7–11.

Mylne, K. and Grahame, N. (2011) Strategic intervention – a novel use of ensembles in forecast guidance. European Meteorology Society Conference 2011 (http:// presentations.copernicus.org/EMS2011-795_presentation.ppt, accessed 19 June 2012.)

Nash, J., Behrens, K. and Leroy, M. (2010) Working to standardize instruments and methods of observation. *WMO Bulletin*, **59**, 18–20.

Parks, K. (2011) Wind Energy Forecasting: What's it Worth? International Conference Energy and Meteorology (ICEM), 8–11 November 2011, Gold Coast, Australia.

Renfrew, I.A., Moore, G.W.K., Kristjánsson, J.E. *et al.* (2008) The Greenland Flow Distortion experiment. *Bull. Am. Meteorol. Soc.*, **88**, 1307–1324.

Richardson, L.F. (1922) *Weather prediction by numerical process*. Republished by Cambridge University Press (2007).

Rodwell, M.J., Richardson, D., Hewson, T.D. and Haiden, T. (2010) A new equitable score suitable for verifying precipitation in numerical weather prediction. *Q.J.R. Meteorol. Soc.*, **136**, 1344–1363.

Rodwell, M.J., Haiden, T. and Richardson, D.S. (2011) Developments in precipitation verification. *ECMWF Newsletter*, **128**, 11–16.

Saith, N. and Slingo, J. (2006) The role of the Madden–Julian Oscillation in the El Niño and Indian Drought of 2002. *Int. J. Clim.*, **26**, 1361–1378.

Schiller, A., Meyers, G. and Smith, N.R. (2009) Taming Australia's Last Frontier. *Bull. Am. Meteorol. Soc.*, **90**, 436–440.

Schmetz, J., Pili, P., Tjemkes, S. *et al.* (2002) An Introduction to Meteosat Second Generation. *Bull. Am. Meteorol. Soc.*, **83**, 977–992.

Stark, J.D., Donlon, C.J., Martin, M.J. and McCulloch, M.E. (2007) OSTIA: An operational, high resolution, real-time, global sea surface temperature analysis system. In: IEEE, *Oceans '07. Marine challenges: coastline to deep sea*. Aberdeen, Scotland. IEEE.

Tiedtke, M. (1989) A comprehensive mass flux scheme for cumulus parametrization in large-scale models. *Mon. Wea. Rev.*, **117**, 1779–1800.

Toth, Z. and Kalnay, E. (1993) Ensemble forecasting at NMC: the generation of perturbations. *Bull. Am. Meteorol. Soc.*, **74**, 2317–2330.

Toth, Z. and Kalnay, E. (1997) Ensemble forecasting at NCEP: the breeding method. *Mon. Wea. Rev.*, **125**, 3297–3318.

Trobec, J. (2011) Role of the broadcaster in communicating forecast uncertainty. http://presentations.copernicus.org/EMS2011-798_presentation.pdf (accessed 20 June 2012).

Wilson, D. R. and Ballard, S.P. (1999) A microphysically based precipitation scheme for the U.K. Meteorological Office Unified Model. *Q.J.R. Meteorol. Soc.*, **131**, 1607–1636.

Woolnough, S.J., Vitart, F. and Balmaseda, M.A. (2007) The role of the ocean in the Madden–Julian Oscillation: implications for MJO prediction. *Q.J.R. Meteorol. Soc.*, **133**, 117–128.

Index

Adjoint *see* Data assimilation
Advanced Microwave Sounding
      Unit (AMSU) 103
Advanced Scatterometer
      (ASCAT) 47
Anelastic approximation *see*
      Approximations in numerical
      models
ARGO float programme 186
Atlantic Hurricane Forecasts 182,
      192, 202
Atmospheric dispersion models 145
Atmospheric Motion Vectors
      (AMVs) 45
Agriculture, forecasts for 1, 165
Aircraft reports 42, 106
Aircraft Meteorological Data Relay
      (AMDAR) 42
Anomaly Correlation Coefficient
      (ACC) 210
Approximations in numerical
      models
   Anelastic 62
   Hydrostatic 62
   Quasi-geostrophic 62–3
   'Shallow Atmosphere' 61
   'Traditional' 61
Australian Bureau of
      Meteorology 4, 157

Bias, numerical model 139, 198, 208
Bjerknes, Jakob 181
Blanford, H.F. 189

Bogussing 105
Boscastle Storm 120
Boundary conditions for regional
      models 115–18, 122–3
Boundary Condition Problem
      (BCP) 180
Breeding methods *see* Initial
      condition perturbations
Brier Score 217–19
Buoy data 37–40, 185–6, 193

Central England Temperature
      (CET) 27
Centrifugal force 57
Charney, Jule 5, 63
Chicago University 5
Civil aviation, forecasts for
      167–71
Climatology 28, 206, 208
Climate change, anthropogenic 192
Cloud condensation nuclei
      (CCN) 75
Clustering 135–6
Colorado State University
      (CSU) 192
Commercial shipping, forecasts
      for 1, 145
Conservation of mass 56, 58
Conservation of water vapour 56,
      60–61
Contingency tables 212–15
Coriolis force 57

Operational Weather Forecasting, First Edition. Peter Inness and Steve Dorling.
© 2013 John Wiley & Sons, Ltd. Published 2013 by John Wiley & Sons, Ltd.

Printed and bound by CPI Group (UK) Ltd, Croydon, CR0 4YY

27/10/2024

14580165-0001